Environment and Ecology

Environment and Ecology

Kamal Kant Joshi
Assistant Professor
Dehradun Institute of Technology
Dehradun, Uttarakhand, India

I.K. International Publishing House Pvt. Ltd.
NEW DELHI • BANGALORE

Published by

I.K. International Pvt. Ltd.
S-25, Green Park Extension
Uphaar Cinema Market
New Delhi–110 016 (India)
E-mail:info@ikinternational.com
Website: www.ikbooks.com

ISBN 978-93-81141-42-7

Reprint 2014

Published by Krishan Makhijani for I.K. International Publishing House Pvt. Ltd., S-25, Green Park Extension, Uphaar Cinema Market, New Delhi–110 016 and Printed by Rekha Printers Pvt. Ltd., Okhla Industrial Area, Phase II, New Delhi–110 020.

Dedicated to
My Parents

PREFACE

This book is deliberated to impart the basic knowledge in Environmental Sciences in accordance with the modern trends. Level of the book covers the syllabi of U.P. Technical University (UPTU), Uttarakhand Technical University (UTU) and also of those being taught at undergraduate standard of engineering colleges and other universities. It also covers a part of syllabi of various competitive and other national level examinations. Contents of the book will be fully helpful to the students appearing for various examinations such as M.Sc. (Environmental Science). M.Tech. (Environmental Engineering) and Life Science for GATE etc. The topics in the book related to various ecosystems, biodiversity, pollution and natural resources etc. are updated ad full of information. This book also emphasizes on some technical point like biodiversity measurement, ETP process, Ozone layer depletion process, acid rain and air pollution control equipment. The book is equipped with several attractive features in the form of objective questions, important days, analysis and discussion on biodiversity conservation and environmental acts. The book is corroborated by a large number of real photographs and diagrams.

I acknowledge the inspiration and blessings of my respected teacher Prof. Dinesh Bhatt, Department of Zoology and Environmental Science, Gurukula Kangri University, Haridwar for his constructive suggestions and valuable ideas during the production of the present work. My special thanks to my teachers Prof. B.D. Joshi and Dr. P.C. Joshi at Department of Zoology and Environmental Sciences, Gurukula Kangri University, Haridwar for their encouragements.

I express my deep love and gratitude to my respected parents and family members Smt. Kiran Joshi (wife), my adorable son Manas, Mr. and Mrs. C.K. Joshi and Mr. and Mrs. Dev Anand Joshi.

I extend special thanks to my Head of Department Prof. Naveen Singhal and colleagues Dr. Anoop Kathait, Dr. Anil Kumar, Dr. Navneet Rai and Mr. Vikas Saini for their help and moral support.

Enormous effort has been made to avoid errors and mistakes; however, their presence cannot be ruled out. Any suggestion to improve this book will be highly appreciated.

—Author

CONTENTS

1

ENVIRONMENT AND ECOLOGY

1.1 INTRODUCTION

Ecology is the study of interaction among organisms and their environments. However, many scientists have described it as the relationship between living things (biotic) and their environment (abiotic). The word ecology derives from two words, *oikos*, meaning "household" and *logos* meaning "Study". Thus, ecology is the study of organisms and all the functional processes that make the house habitable. In this way, ecology is the study of "life at home". In other words, "the relation between organisms and their environment is ecology.

The term ecology was introduced by the German biologist Ernest Hackel in 1869. Hackel defined ecology as "*the study of the natural environment including the relations of organisms to one another and their surroundings*". Therefore, scientists and researchers have changed some words in this definition, although the meaning is same.

1.2 ENVIRONMENT

The word "environment" means surroundings. Environment Protection Act (1986) defined environment as "the sum total of water, air and land, their interrelationship among themselves and human beings and other living beings". Many environmentalists defined environment as the surrounding area composed of atmosphere, hydrosphere, and lithosphere and biotic-abiotic factors.

Atmosphere : Atmosphere is the gaseous part of the environment, or in other words, the gaseous section of the environment is called atmosphere. Atmosphere consists of different layers: Troposphere, Stratosphere, Mesosphere, Ionosphere and Exosphere.

Lithosphere : Lithosphere (earth's crust and upper mantle) is the solid part of the earth like rocks, solids. The solid composition section of environment is called lithosphere in which animals and plants survive.

Hydrosphere : Hydrosphere is our surrounding liquid section of environment (groundwater, surface water, ocean water). Hydrosphere provides suitable habitat for aquatic life.

1.2.1 Biotic Components

Biotic components are living organisms that survive in a healthy environment. It includes plants, animals and microbes. Among these components, only green plants receive energy from the sun and utilize this energy to produce food in the form of carbohydrates, proteins and fats. Animals and microbes obtain their food from plants.

1.2.2 Abiotic Components

The abiotic components are non-living and include climatic and edaphic factors. The climatic factors include temperature, humidity, rain, snowfall, etc. The edaphic factors comprise the soil. A brief introduction of abiotic factors is given as follows:

Temperature : Temperature is an abiotic factor. It affects the environment and living organisms. This is due to this factor that we can see a variety of plant and animal species in nature.

Light : Light is another essential abiotic factor for plants. Plants show various adaptations for obtaining optimum light. The activities of a large number of organisms, e.g., leech, moths are regulated by light.

Wind : Wind is an abiotic factor that helps in the dispersal of seeds and fruits of many plants. It also plays an important role in species variation and distribution.

Water : Water is an essential requirement of life and abiotic factor. Water provides a unique habitat and distributes the organisms, e.g., salty water (marine) species survive in marine water, and river water habitat species survive in sweet water (river).

These living and non-living components function interdependently as ecological systems or ecosystem. Energy flowing from the sun and various biochemical processes enable the ecosystems to function. Biochemical processes cycle nutrients, involving carbon, oxygen, nitrogen, phosphorus and sulphur besides minerals like iron, copper etc. Naturally, ecosystem would not function if the energy flow were to slow down or stop. And, if ecosystem were to malfunction, the integrity of all biotic communities would be in peril.

It is evident that human beings, environment and the earth system (lithosphere, hydrosphere, atmosphere, and biosphere) are inseparably and interdependently linked. If one is in distress, the other two are also affected (Fig. 1.1).

1.3 STATUS OF ENVIRONMENTAL EDUCATION

We all are associated with our nature/environment directly or indirectly. However, we ignore this fact. To keep in mind the importance of environment and sustainable development for future, the Government of India, Human Resources and Development (UGC, AICTE) have decided to make the study of environment education a compulsory subject in schools and colleges. On the other hand, many MNCs and national companies sign up such fresher students who are aware of environment and nature education, so that they cannot only improve the company's growth but also our environment. The

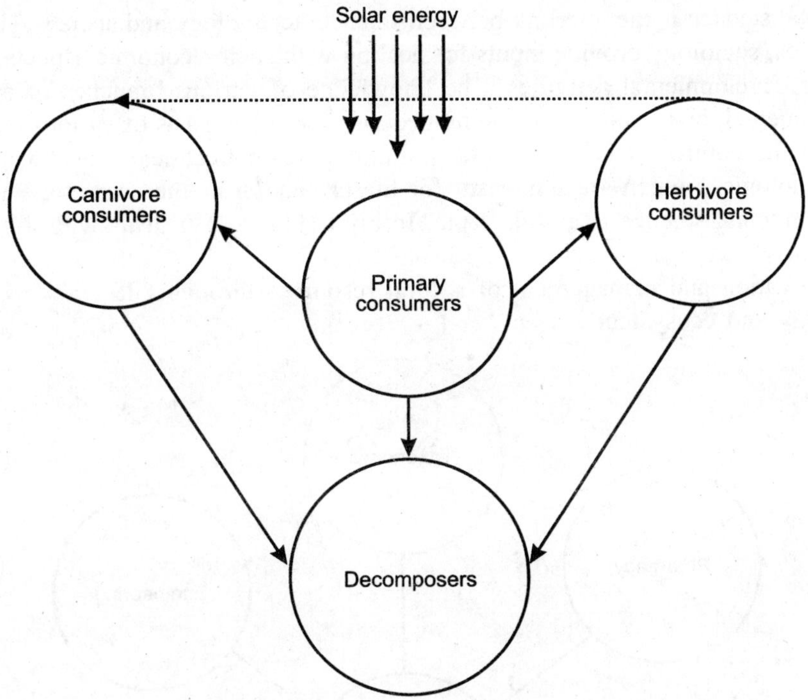

Fig. 1.1: Humans, the environment and the earth system are inextricably linked and function interdependently.

prescribed curriculum will not help a student in getting a mound of rubbish outside a house or school cleared. Students are being taught mere facts and figures. We have to make our students realize that they are a part of the problem, and therefore they have to be the part of the solution.

NASA (USA) has started the Global Learning and Observations to Benefit the Environment (GLOBE) programme in about 50,000 schools all over the world. Out of these, 86 are in India, and are enrolled in the GLOBE programme. A school in Lucknow has set up a small weather monitoring station in its campus. Children use the station to maintain temperature and cloud charts, measure rainfall, gather weather-related information and feed it into the GLOBE website. The data is then used to forecast worldwide weather trends and to develop environment protocols.

1.4 MULTIDISCIPLINARY NATURE OF ENVIRONMENTAL SCIENCE

Environmental Science/Ecology is the branch of biology that has emerged from biology as an essentially new integrative discipline, which links physical and biological processes and forms a bridge between the natural science and social science. In other words, the environmental science is a multidisciplinary subject studied in technical and science subjects.

Environmental science is the interlink between science, technology and society. The subjects like economics, sociology provide inputs for dealing with socio-economic aspects associated with various developmental activities. The knowledge of various branches of engineering (i.e., environmental, civil, and chemical engineering) forms the basis of various technologies dealing with the control of environmental pollution, waste treatment and development of cleaner technologies, which are important for the protection of the environment.

Environmental science is a wide topic. In brief (Fig. 1.2) It deals with the following disciplines:

- Conservation and management of natural resources through GIS
- Ecology and ecosystem

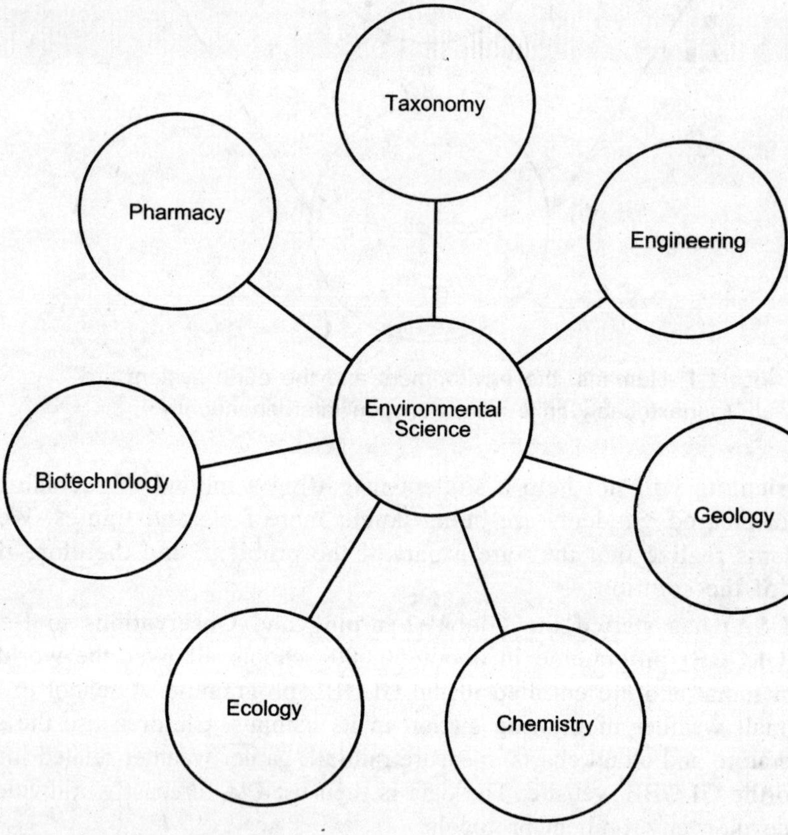

Fig. 1.2: The relationship of environmental science with other branches.

- Environmental pollution and control
- Natural disasters and their control
- Research and development
- Social issues
- Biodiversity assessment
- Geology

EXERCISE

QUESTIONS

1. What is the difference between environment and ecology?
2. Discuss: environmental science is a multidisciplinary subject.
3. How does environmental studies play an important role in technical education?
4. What are the environmental factors? Give their details.
5. How are human beings directly linked with the environment?

OBJECTIVE TYPE QUESTIONS

1. Environmental factors are:
   (a) biotic, abiotic (b) living, non-living
   (c) both (d) wood and animal
2. Biotic factors are:
   (a) plant, animals and microbes (b) man and animal
   (c) wind & water (d) none
3. Environmental science subject is:
   (a) scientific (b) non-scientific
   (c) technical (d) science and social subject
4. Who proposed the term ecology?
   (a) E. Hackel (b) Thomson
   (c) A.G. Tansley (d) None
5. Atmosphere is composed of:
   (a) troposphere (b) stratosphere
   (c) mesosphere (d) all
6. Environmental Protection Act was proposed in:
   (a) 1986 (b) 1896
   (c) 1996 (d) 1869
7. A main source of energy in environment is:
   (a) coal (b) LPG
   (c) sun (d) wind
8. Greek '*oikos*' means:
   (a) plant, animals and microbes (b) man and animal
   (c) household (d) none
9. The largest amount of usable water found on earth is used for:
   (a) industrial uses (b) irrigation
   (c) household (d) crop irrigation

10. The gas in the atmosphere that can absorb the UV radiation is:
(a) O_3 (b) SO_2
(c) CO_2 (d) none

11. GIS means:
(a) geographic information system (b) geographic information satellite
(c) both (d) geographic institute for studies

12. Environmental Studies is:
(a) plant & animals (b) man & nature
(c) wind & water (d) development with nature

13. Pedology is the branch of science dealing with:
(a) soil (b) rocks
(c) water (d) animal

14. Lithosphere is related to:
(a) land & soil (b) water & soil
(c) wind & water (d) none

15. Hydrosphere is related to:
(a) water & soil (b) living, non-living
(c) water (d) all

16. Atmosphere is a part of:
(a) gases (b) water
(c) wind (d) none

17. The branch of science in which we study about the nature and its components is:
(a) ecology (b) environmental science
(c) both (d) none

18. Snowfall is a:
(a) biotic factor (b) abiotic factor
(c) both (d) none

19. Edaphic factors are:
(a) soil (b) water
(c) wood (d) all

20. Microbes are a:
(a) biotic factor (b) abiotic factor
(c) edaphic factor (d) none

ANSWERS

1. (a)	2. (a)	3. (d)	4. (c)	5. (d)	6. (a)	7. (c)	8. (c)	9. (d)	10. (a)
11. (b)	12. (d)	13. (a)	14. (a)	15. (c)	16. (a)	17. (b)	18. (b)	19. (d)	20. (a)

2

NATURAL RESOURCES

2.1 INTRODUCTION

The word 'resources' means a source of supply or support that is generally held in reserve. In other words, natural resources are the materials, which living organisms can take from nature for their sustenance, or any component of the natural environment that can be utilized by man to promote his welfare. Natural resources are the components of atmosphere, hydrosphere and lithosphere, which are useful and basic needs of life. These include energy, air, water, minerals, plants, animals and soil. For human beings resources are defined as those substances which are required for survival, comfort and prosperity. They are obtained directly from the environment. Some resources are likely to get exhausted soon, whereas others would be lasting for a long period of time. Thus, depending upon the availability of resources during our continuous use, resources are divided into two categories.

1. **Renewable Resources**: Renewable resources are such materials which can be renewed in a short time, in other words, such natural resources renew or redevelop in a short period. These include sunlight, plants, animals, soil and water.
2. **Non-renewable resources**: Non-renewable sources are minerals, fossil fuels, and non-material sources, which are present in a limited amount in the nature, and take a long time to renew.

   On the basis of origin, they are classified into the following two categories:

(a) **Organic resources**: These are the resources obtained from biosphere, for example, forest and forest products, crops, birds, animals, fish and other marine forms, coal, mineral oils etc.

(b) **Inorganic resources**: These are the resources composed of inorganic matter, also called abiotic resources e.g., land, water etc.

2.2 FOREST RESOURCES

Forest is the major component of environment, which provides protection to wild flora and fauna, help in balancing the gaseous cycle of atmosphere, tends to increase local precipitation and water holding capacity of soil, and maintains soil fertility. On the basis of forest survey report, India have about 6,33,397 km^2 forest cover area. This is about 19.27% of India's total geographical area. India is endowed with diverse forest types like tropical wet evergreen forest in the northeastern to the tropical thorn forest in the central and western India. The forest of the India can be classified into 16 major groups and 221 types. The four major types of forest in India and their climatic conditions are given in Table 2.1.

Table 2.1: Climatic conditions and major forest types of India.

Tropical forests	Annual Temp. (°C) 23° – 37°	Annual rainfall (mm) 2000-3500
Tropical wet evergreen Tropical semi-evergreen Tropical moist deciduous Tropical littoral and swamp Tropical dry deciduous Tropical thorn Tropical evergreen		
Subtropical forests	Annual Temp. (°C) 22° – 23°	Annual rainfall (mm) 900-1600
Subtropical broad leaved hill forest Subtropical pine Subtropical dry evergreen		
Temperate forests	Annual Temp. (°C) 6° – 20°	Annual rainfall (mm) 1000-2500
Montane wet temperate Himalayan dry temperate Himalayan moist temperate		
Sub-alpine and alpine forest	Annual Temp. (°C) 0° – 15°	Annual rainfall (mm) 500-1700
Sub-alpine Moist alpine shrub Dry alpine shrub		

2.2.1 Use and Overexploitation of Forest Resources

A forest is a biotic community predominantly consisting of trees, shrubs and other woody vegetations, usually with a closed canopy. This invaluable renewable nature resource is very useful in several ways. Their benefits can be studies in two types:

2.2.1.1 *Direct benefits (productive function) from forests*

(a) Fuel wood: Dry shoots and leaves are used as fuel in many parts of India. It is an alternate source of energy and used for cooking purpose.

(b) **Timber**: Heavy shoots or stems are used as timber for making furniture, tool handles, railways sleepers, boats etc.

(c) **Food**: Forest provides food and shelter to wild species and plays an important role in nature ecosystem.

(d) **Forest products**: Tannins, gum, drugs, species, insecticides, waxes, honey, musk etc. are the forest products utilized by human beings in several ways like medicine, cosmetic etc. and they also help to increase the revenue of government.

2.2.1.2 *Indirect benefits (protective function) from forest*

(a) **Conservation of soil**: A good vegetation cover in forests prevents soil erosion by binding the soil with the network of roots of plants and trees and reduces the velocity of wind and soil. The green belt along the road also helps in controlling landslides.

(b) **Soil improvement**: Forest is much fertile compared to other land. This may be due to rich litter and foliage.

(c) **Reduction of atmospheric pollution**: Forest vegetation helps to reduce air pollution and temperature due to photosynthesis. A good forest cover increases the atmospheric humidity which affects rainfall and cools the temperature of atmosphere.

2.3 THREATS TO FOREST

2.3.1 Deforestation

Deforestation is the permanent destruction of forest and woodlands. Deforestation has resulted in the reduction of indigenous forests to 4/5 of their pre-agriculture land. Indigenous forests now cover 19.6% of the earth's surface. The difference between forest and woodland is that in a forest, the crowns of individual trees touch to form a single canopy, but in a woodland, trees grow far apart so that the canopy is open. Currently on the basis of forest survey report, 12 million hac. of forests are cleared annually. Almost all of this deforestation occurs in the moist and tropical forests. At this rate, all moist tropical forests could be lost by the year 2050.

The destruction of forests due to unscrupulous and indiscriminate felling of trees has led to an over all deterioration of our environment and is posing a serious threat to the quality of life in future.

2.3.1.1 *Causes of deforestation*

There are several causes which promote deforestation:

(a) **Population explosion**: Population explosion is the main threat to the environmental problems. Vast areas of forest land are cleared of trees to reclaim land for human settlements (factories, agriculture, housing, roads, etc.). Growth of population increases the demand for forest products like timber, fire wood, paper and other valuable products of industrial importance, all necessitating the felling of trees.

2.3.2 Forest Fire

The most common hazard in forests is forests fire. Forest fires are as old as the forests themselves. They pose a threat not only to the forest wealth, but also to the entire regime of fauna and flora, seriously disturbing the biodiversity and ecology and environment of a region. During summer, when there is no rain for months, the forests become littered with dry senescent leaves and twinges, which could burst into flames ignited by the slightest spark. About 90% of the forest fires in India are created by humans. The normal fire season in India is from the month of February to mid June. The Himalayan forests, particularly, Garhwal Himalayas, have been burning regularly during the last few summers, with colossal loss of vegetation cover of that region.

2.3.2.1 *Causes of forest fire*

On the basis of forest fires sources, it can be studied as natural and man-made.

(a) **Natural causes**: Many forest fires start from natural causes such as lightning, which sets trees on fire. However, rain extinguishes such fires without causing much damage. High atmospheric temperatures and dryness (low humidity) offer favourable circumstances for a fire to start (Fig. 2.1).

(b) **Man-made causes**: Fire is caused when a source of fire like naked flame, cigarette or bidi, electric spark or any source of ignition comes into contact with inflammable material (Fig. 2.2).

2.3.2.2 *Classification of forest fire*

Forest fire can be classified into three categories:

- Natural or controlled forest fire.

Fig. 2.1: Forest fire in tropical dry deciduous forest.

Fig. 2.2: Forest fire control in India using latest technologies and fire-proof safety clothing.

- Forest fires caused by heat generated in the litter and other biomes in summer through carelessness of people (human neglect)
- Forest fires purposely caused by local inhabitants.

2.3.2.3 *Types of forest fire*

There are two types of forest fires:

(a) **Surface fire**: A forest fire may burn primarily as a surface fire, spreading along the ground, as the surface litter (senescent leaves and twigs and dry grasses etc.) on the forest floor is engulfed by the spreading flames (Fig. 2.3a).

Fig. 2.3 (a): Surface fire in a forest.

(b) **Crown fire**: The other type of forest fire is crown fire in which the crown of trees and shrubs burn, often sustained by a surface fire. Crown fire is particularly dangerous in a coniferous forest because resinous materials given off during burning logs burn furiously. On hill slopes, if the fire starts downhill, it spreads up fast as heated air adjacent to a slope tends to flow up the slope, spreading flames along with it.

Fig. 2.3 (b): Crown fire in a forest.

2.3.2.4 *Impacts of forest fires on biological environment*

Forest fires are also cause serious health hazards by producing smoke and noxious gases. The burning of vegetation gives off not only carbon dioxide but also a host of other greenhouse gases such as carbon monoxide, methane, hydrocarbons, nitric oxide and nitrous oxide, which lead to global warming and ozone layer depletion. Consequently, thousands of people suffered from serious respiratory problems due to these toxic gases. Burning forests and grasslands also add to already serious threat of global warming. Recent measurement suggests that biomass burning may be a significant global source of methyl bromide, which is an ozone depleting chemical.

2.3.3 Grazing Animals

Sometimes villagers (living adage of forest land) free their cattle for grazing into the forest. During grazing, these cattle destroy the humus (soil upper layer) and herbs of soil by their hooves. In this way, soil erosion and desertification of the previously fertile forest area results. (Fig. 2.4).

Fig. 2.4: Cattle grazing in a forest.

2.3.4 Shifting Cultivation

Shifting cultivation is common mostly in tropical countries. In India, the people of eastern and north-eastern region (Assam, Arunachal Pradesh, Meghalaya, and Manipur) practice shifting cultivation on hill slopes. Eighty-five per cent of the total cultivation in northeast India is by shifting cultivation. The current practice of shifting cultivation in India is an extravagant and unscientific form of land use. The evil effects of shifting cultivation are devastating and far-reaching in degrading the environment and ecology of these regions.

Frequent shifting from one land to the other has affected the ecology of these regions. The area under natural forest has declined; the fragmentation of habitat, local disappearance of native species and invasion by exotic weeds and other plants are some of the other ecological consequences of shifting agriculture. The earlier 15–20 year cycle of shifting cultivation on a particular land has reduced to 2–3 years now. This has resulted in large-scale deforestation, soil and nutrient loss, and invasion by weeds and other species. The indigenous biodiversity has been affected to a large extent (Figs. 2.5 and 2.6)

Fig. 2.5: Shifting cultivators use fire for crop production.

Fig. 2.6: Shifting cultivators removing the trees for farming.

2.3.4.1 *Impact of shifting cultivation*

- Burning of the vegetation. Small, cut-trunks portion and roots are normally not removed. The herbs, shrubs and twigs and branches (slashed vegetation) are burnt.
- Sowing of seeds, by dibbling, generally of cereals, vegetables and oil seeds.
- Continuing cultivation for a few years.

- Abandoning the cultivated site and shifting to other forest sites.
- Returning to the former site, and once again practice shifting cultivation on it.

Effects of Deforestation: Forests are closely related with climate change and biological diversity. Large scale deforestation has many far reaching consequences:

(a) Habitat destruction of wild animals.
(b) Increase the soil erosion due to reduction of vegetation cover.
(c) Decrease in availability of forest products.
(d) Loss of culture diversity.
(e) Deforestation reduces the ground level water.
(f) Rise in CO_2 level has resulted in increased thermal level of the earth, which in turn results in melting of ice caps and glaciers and consequent flooding of coastal areas.
(g) Decrease the oxygen level of the deforestation land.
(h) Increase the temperature and reduce the precipitation.

2.4 WATER RESOURCE

Water resources are sources of water (H_2O) that are potentially useful to humans. Uses of water include agricultural, industrial, household, recreational and environmental activities. Ninety-seven per cent of water on the earth is salt water and rest 3% is freshwater, of which slightly over two-thirds is frozen in glaciers and polar ice caps, while the remaining unfrozen freshwater is mainly found as groundwater, with only a small fraction present above the ground or in the air.

India receives annual precipitation of about 4000 km^3. Although India occupies only 3.29 million km^2 geographical areas, which forms 2.4% of the world's land area, it supports over 15% of the world's population. The population of India as on 1 March 2001 stood at 1,027,015,247 persons. Thus, India supports about 1/6th of world population, 1/50th of world's land, and 1/25th of world's water resources. The total utilizable water resources of the country are assessed about 1086 km^3.

2.4.1 Sources and Use of Freshwater

2.4.1.1 *Surface water*

Surface water sources are sea, rivers, lakes or freshwater wetlands. Surface water is naturally replenished by precipitation and naturally lost through discharge to the oceans, evaporation, and sub-surface seepage. Although the only natural input to any surface water system is precipitation within its watershed, the total quantity of water in that system at any given time is also dependent on many other factors. These factors include storage capacity in lakes, wetlands and artificial, the permeability of the soil beneath these storage bodies, the runoff characteristics of the land in the watershed, the timing of the precipitation and local evaporation rates. All of these factors also affect the proportions of water lost.

2.4.1.2 *River*

The water coming through precipitation (rainfall, snow) gets stored into the ice caps and glaciers. These glaciers are origin of many rivers, however, during rain many streams become full of water and known as rain fed river. Sometimes rainwater does not percolate into the ground or is not returned to the atmosphere as evaporation. It gets stored into the form of streams, lakes, ponds, wetlands or artificial reservoirs, which is known as surface water. The surface water is largely used for irrigation, industrial use, public water supply, navigation, etc. A country's economy largely depends upon its rivers.

2.4.1.3 *Artificial and natural reservoirs*

These are formed by the construction of dams across river valleys. However, natural reservoirs known as lakes are generally small, temporary or permanent shallow water bodies. As their water is much more accessible than the groundwater or glaciers, they are considered an important source of freshwater supply.

2.4.1.4 *Sea water*

Sea is a mixture of various salt and water. A huge amount (97%) of the surface water on the earth is salty. This ocean water cannot be used openly however to solve the water crisis in coastal areas this water is used after distillation.

2.4.1.5 *Groundwater*

Groundwater is an important natural resource. About 9.86% of the total freshwater resources are in the form of groundwater, and it is about 35-50 times that of surface water supplies (Fig. 2.7) About 1/3 of the state's population and 90% of the state's rural population rely on groundwater for drinking. Groundwater is the water found in the subsurface, and can be found just a few inches below the ground surface during rainy periods to hundreds of feet in the subsurface. Groundwater fills the pores between the grains of the subsurface materials. Porosity defines how much water fills the pores in geologic materials, and varies from less than 5% for some bedrocks to over 40% for sand and gravels. Sand and gravels tend to have high hydraulic conductivity (0.1 cm/sec), while fine grained materials like silts and clays have much lower hydraulic conductivity (0.000001 cm/sec).

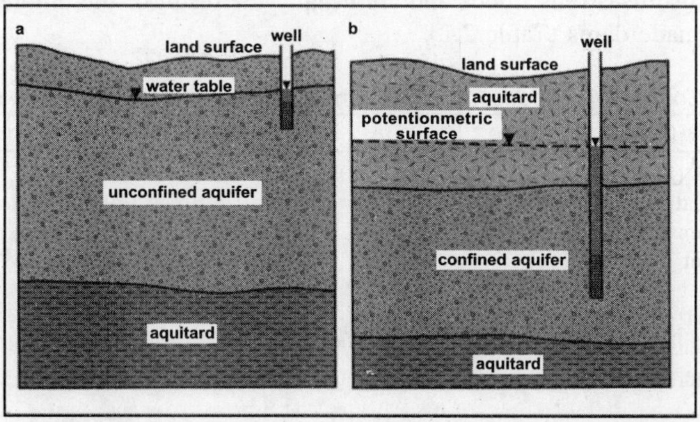

Fig. 2.7: Groundwater level in different layers of the earth.

2.5 FLOOD

The term "flood" is a general or temporary condition of partial or complete inundation of normally dry land areas from overflow of inland or tidal water or from the unusual and rapid accumulation or runoff of surface waters from any source. The flood causes loss of human life and widespread damage to property. Unimaginable damage to agriculture takes place affecting the planning and upsets the financial budgeting, thereby slowing down the whole economy of the country. Majority of countries do not document or map floods methodically. Water remains stagnant after the flood recedes, sources of drinking water get polluted and the foods get spoiled. The only way to fight the floods is to try to predict the flood-prone area, prepare for it, and educate the people.

2.5.1 Types of Floods in India

Floods of different types occur in India. These include the rainfall, rainstorm, coastal and dam-failure floods. Snowmelt floods are not very important in the Indian context.

2.5.1.1 *Rainfall floods*

These are frequent in the annual flood series for most rivers of the subcontinent. Rainfall floods result from heavy precipitation in association with active to vigorous monsoon conditions for a number of days.

2.5.1.2 *Rainstorm floods*

These are generated by excessively heavy rainfall associated with low depressions and cyclonic storms originating over the Bay of Bengal or the adjoining coastal belt. Cyclonic storms contribute significantly to the total precipitation and produce devastating floods.

2.5.1.3 *Coastal floods*

These are connected with storm surges and occur in some parts of the eastern coast of India and are produced by severe cyclonic storms originating over the Bay of Bengal.

2.5.1.4 *Dam-failure floods*

These are the most disastrous floods and are usually the result of the failure or breaching of natural or man-made dams (Table 2.2)

Table 2.2: Some major floods caused by the failure of man-made dams (CSE, 1991)

Name of Dam	Dam failure flood year
Tigra (M.P.)	1971
Kundali (Maharashtra)	1925
Kadam (Andhra Pradesh)	1958
Panset (Maharashtra)	1961
Kharagpur (Bihar)	1961
Morvi (Gujarat)	1979
Jaswant Nagar (Rajasthan)	1979
Chandora (M.P.)	1991

2.5.2 Causes of Floods

Heavy downpour in the form of rain brings down more water than can be disposed of by combined factors, natural and man-made systems, causes flooding. The rivers overflow embankments may be breached. Generally heavy rains due to the storm and hurricane bring unmanageable amount of water causing flash floods. However, most floods follow heavy rain, or rapid widespread melting of deep snow. Flood forecasters also have to worry about several other factors. Here are some factors that help determine whether a flood is a minor or major.

2.5.2.1 *Deep snow cover*

Deep snow can melt into a lot of water. Deep snow very rarely causes flooding by itself. Often, heavy rain and rapid warm up combine with rapidly melting snow to cause major flooding problems. For example, Fargo, S.D. had a 15 inch snow pack at the beginning of March 1994.

2.5.2.2 *Frozen ground*

Frozen soil cannot absorb as much water as unfrozen soil. Rain or rapid snow melt atop frozen soil can cause flooding that would not have occurred if the soil were not frozen.

2.5.2.3 *Wet or saturated soil*

Saturated soil cannot absorb rainwater. The excess water becomes runoff that rapidly flows into rivers and streams. Unsaturated soil acts like a sponge, absorbing some of the water from rain or snow melt. Saturated soil by itself does not cause flooding. Usually, heavy rain or rapid snow melt combined with saturated soil causes the flooding.

2.5.2.4 *Reservoirs*

Reservoirs are large, mostly man-made basins which build during dam construction that hold water for irrigation and drinking proposes. Reservoirs can alleviate river flooding by absorbing and spreading out flood crests flowing down the river. This would reduce the height in which the water rises downstream of the reservoir. If the reservoir is already full, it cannot absorb any water from swollen rivers.

2.5.2.5 *High river and stream levels*

Streams or rivers that are already at bankfull can be a precursor to major flooding. Heavy rain or rapid snow melt that flows into an already full river will cause the river to overflow its banks and flood nearby locations. High river levels, such as those in the Ohio Valley in the spring of 1997 make forecasters very nervous anytime a storm threatens to dump heavy rain over the region. A prolonged dry spell, however, can alleviate flooding concerns.

2.5.2.6 *Deforestation*

This is perhaps the most important and influential factor of them all. Due to the heavy rainfall in deforestation land, increase the flow of water with silt or soil and create the flooding condition in river.

2.5.3 Impact of Flood

- The large amount of pooled water remaining after the flood leads to an increase in mosquito populations, and cause mosquito-borne diseases, i.e., malaria, dengue, and other diseases like typhoid and cholera, etc.
- Due to the flood, many wild animals get displaced from their natural habitats, and many domestic animals are also rendered homeless.
- Flood water may move hazardous chemicals and pollute with the chemicals. A rich amount of water particulate underground may get polluted in this way. Flood area becomes chemically polluted and spreads water-borne diseases.
- Flood affects crops, cropland and human settlements (Fig. 2.8 (a)). Deposition of thick silt and slush in croplands makes it barren and spreads the water-borne diseases in urban areas.
- Floods damage the residential properties, transportation and utilities as well as affect economy (Fig. 2.8(b)).

Fig. 2.8. (a) Showing submerged village with flood.

Fig. 2.8. (b) Flood in urban areas.

2.6 DAMS

Dam is a physical barrier constructed across the river or waterway to control the flow level of water. Dam may be constructed for flood control, irrigation needs, hydroelectric power production, and recreation usage (Fig. 2.9).

Fig. 2.9: Tehri dam in Uttarakhand.

2.6.1 Types of Dams

There are several types of dams including earth fill, rock fill, gravel fill, solid-masonry gravity, solid masonry arch, arch gravity, structural masonry and steel or timber. The choice of dam is decided upon by examining foundation conditions, load strains, temperature and pressure changes, chemical characteristics of groundwater and possible seismic activity. The most common types of dams are embankment (earth fill) and masonry (concrete) *Embankment* dams retain water across wide rivers when large amounts of fill are needed. They are less expensive and more adaptable to natural changes. *Masonry* dams are made of concrete, and they use the force of gravity and the weight of the materials to counteract the horizontal force of the moving water. Another type of masonry dam is the arch. It consists of a convex arch that faces the reservoir and is reinforced by the pressure of the water.

2.6.2 Benefits of Dams

Water is essential for sustenance of all forms of life on the earth. It is not evenly distributed all over the world and even its availability at the same locations is not uniform over the year. While the maximum parts of the world, which are scarce of water. No doubt the rivers are a great gift of nature and have been playing a significant role in the evolution of various civilizations, nonetheless on many occasions, rivers, at the time of floods, play havoc with the life and property of the people. Management of river waters has been, therefore, one of the most prime issues under consideration. Optimal management of river water resources demands that specific plans should be evolved for various river basins, which are found to be technically feasible and economically viable after carrying out extensive surveys. Since the advent of civilization, man has been constructing dams and reservoirs for storing surplus river waters available during wet periods and for utilization of the same during lean periods. The dams and reservoirs world over have been playing dual role of harnessing the river waters for accelerating socio-economic growth and mitigating the miseries of a large population of the world suffering from the vagaries of floods and droughts. Dams and reservoirs contribute significantly in fulfilling the following basic human needs:

- **Water for drinking and industrial use**: Due to large variations in hydrological cycle, dams and reservoirs are required to be constructed to store water during periods of surplus water availability and conserve the same for utilization during lean periods when the water availability is scarce. Properly designed and well-constructed dams play a great role in optimally meeting the drinking water requirements of the people. Water stored in reservoirs is also used vastly for meeting industrial needs. Regulated flow of water from reservoirs helps in diluting harmful dissolved substances in river waters during lean periods by supplementing low inflows, and thus helps in maintaining and preserving the quality of water within the safe limits.
- **Irrigation**: Dams and reservoirs are constructed to store surplus waters during wet periods, which can be used for irrigating arid lands. One of the major benefits of dams and reservoirs is that water flows can be regulated as per agricultural requirements of various regions over the year. It is estimated that 80% of additional food production by the year 2025 would be available from the irrigation made possible by dams and reservoirs. Dams and reservoirs are most needed for meeting irrigation requirements developing countries, large parts of which are arid zones. There is a need for construction of more reservoir based projects despite widespread measures developed to conserve water through other improvements in irrigation technology.
- **Flood control**: Floods in the rivers have been, many a times playing havoc with the life and property of the people. Dams and reservoirs can be effectively used to control floods by regulating river water flows downstream the dam. The dams are designed, constructed and operated as per a specific plan for routing floods through the basin without any damage to life and property of the people. The water conserved by means of dams and reservoirs at the time of floods can be utilized for meeting irrigation and drinking water requirements and hydropower generation.
- **Hydropower generation**: Energy plays an important role in socio-economic development of a country. Hydropower is a cheap, clean and renewable source of energy. Hydropower is the most advanced and economically feasible resource of energy. Reservoir based hydroelectric projects provide much needed peaking power to the grid. However, large hydro potentials can be exploited through mega hydroelectric projects for meeting power needs on regional or national basis. Small hydro potentials can be exploited through mini/microhydel projects for meeting local power needs of small areas. Besides hydropower generation, multipurpose hydroelectric projects have the benefit of meeting irrigation and drinking water requirements and controlling floods, etc.
- **Recreation**: In the areas where natural surface water is scarce or non-existent, the reservoirs may be a great source of recreation. Reservoirs not only complete their objectives, but also supply recreational benefits such as boating, swimming, fishing and water games.
- **Benefit for fish species**: The reservoirs preserve a rich variety of fish species. The ichthyofauna of a reservoir basically represents the faunal diversity of the parent river system. On the basis of studies conducted so far, large reservoirs, on an average, harbor 60 species of fishes, of which at least 40 contribute to the commercial fisheries. The fast growing Indo-Gangetic carps, popularly known as Indian major carps, occupy a prominent place among the commercially important fishes. (The Indian major carps:

Labeo rohita, L. calbasu, L. fimbriatus, Cirrhinus mrigala, Catla catla, The mahseers: *Tor tor, T. putitora, T. khudree, Acrossocheilus hexagonolepis.)*

2.6.3 Disadvantages of Dams

While dams provide a yeoman service to the mankind, the following impacts of the construction of dams are required to be handled carefully:

- **Resettlement and Rehabilitation**: During the dam construction, rehabilitation is a great problem. Many of the displaced people are not recognized and therefore are not resettled or compensated. Due to the rehabilitation, local communities and their culture are completely removed from the dam site, which has a effect on emotional attachment memories. Affected people living near the reservoir as well as displaced people and downstream communities have often faced adverse health and livelihood outcomes from environmental changes and social disruption.
- **Environment and forests**: Dams contribute to changes in the earth's climate. This is because dams generate methane, a greenhouse gas. Methane is emitted from reservoirs that are stratified, in which the bottom layers are anoxic (i.e., they lack oxygen), leading to degradation of biomass through anaerobic processes. Dam construction acquires a huge amount of land (forest, agriculture land etc) and reservoirs submerge the forest land, vegetation and wildlife habitat.
- **Socio-economic issues**: Millions of people living downstream dams (particularly those who depend on natural flood plain function and fisheries) have also suffered serious harm to their livelihoods, and the future productivity of their resources has been put at risk.
- **Water-borne diseases**: The relationship between the water quality and health is more complicated and only a few water-related problems have been evaluated so far. Human health can be affected by reservoir water in several ways. During construction, several pollutants pollute the reservoir water, huge waterbody rise the mosquito population and these vectors may be the cause of diseases and other water-borne vector (like snails) infectious diseases caused by microorganisms through water.
- **Aquatic ecosystem**: Reservoir is a man-made aquatic ecosystem in nature. In the reservoirs exist riverine and lacustrine characteristics, which depend on the temporal and spatial variations of certain habitat variables. For example, the lotic sector of the reservoir sustains a fluviatile biocoenos, whereas the lentic zone and the bays harbour lentic communities. During the rainy months, when heavy inflow and outflow are found, the whole reservoir imitates a lotic environment, whereas in summer months, when the inflow into and outflow from the reservoir decrease, a lentic condition exists in it. In this way, due to sudden changes in lentic to lotic environment and vice versa, a number of organisms die, some migrate to other surrounding areas, and many adapt to the changed habitat.
- **Sedimentary issues**: Rivers carry different types of sediment, which get down their riverbeds, promote the formation of fertile river banks, river delta, alluvial fans, braided rivers, oxbow lakes and coastal shores. The construction of a dam blocks the flow of sediments in downstream, decrease the fertility of downstream riverbanks and increase

the sediment build-up in the reservoir. However, the rate of sedimentation varies for each dam and river, therefore reservoir water-storage capacity reduces due to the exchange of storage space into the sediment. Diminished water storage capacity decreases the hydropower production, reduces the availability of water for irrigation, and ultimately results in the expiration of the dam and river.

2.7 RAINWATER HARVESTING

Rain is the primary source of water for humans. Rivers, lakes and groundwater are all secondary sources of water. In present times, we depend entirely on such secondary sources of water. In India, almost 90% of drinking water is supplied in rural areas through groundwater. Groundwater is also widespread in its occurrence and is easily available without any restrictions on its use. However, in some areas, there is a significant decline in the underground water levels and the quality of this water source has been adversely affected by various activities. The most important remedial measure to arrest the decline trend of groundwater level and the quality is to take up rainwater harvesting and recharge of groundwater.

Rainwater harvesting means capturing/storing the rain where, or capturing the run-off and recharging the underground water in villages or towns (Fig. 2.10) The rainwater harvesting technique can serve to provide drinking water, irrigation water and increase ground level water recharge, reduce storm water discharges, urban floods and overloading of sewage treatment plants and reduce seawater ingress in coastal areas. Rainwater harvesting is the total amount of water that is received in the form of rainfall (*rainwater endowment)* of the area. Out of this, the amount that can be effectively harvested is called the *water harvesting potential*.

Water harvesting potential = Rainfall (mm) × Collection efficiency

The *collection efficiency* accounts for the fact that all the rainwater falling over an area cannot be effectively harvested because of evaporation, spillage etc. Factors like runoff coefficient and the first-flush wastage are taken into account while estimating the collection efficiency.

For example, if we have a flat terrace area of 100 sq m (floor) Assume the average annual rainfall in this area is approximately 600 mm. In simple terms, this means that if the terrace floor is assumed to be impermeable, and all the rain that falls on it is retained without evaporation, then, in one year, there will be rainwater on the terrace floor to a height of 600 mm.

- Area of plot = 100 sq. m. (120 square yards)
- Height of the rainfall = 0.6 m (600 mm or 24 inches)
- Volume of rainfall over the plot = Area of plot × height of rainfall
- Assuming that only 60 per cent of the total rainfall is effectively harvested
- Volume of water harvested = 36,000 litres (60,000 litres × 0.6)

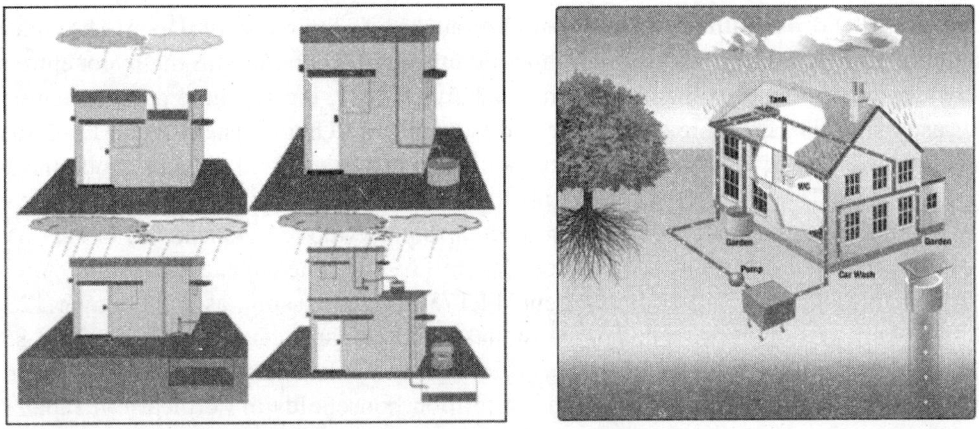

Fig. 2.10: Rainwater harvesting.

2.8 CASE STUDY

2.8.1 Hurricane Katrina

The essential ingredients of hurricane are warm water and moist warm air. Hurricanes start when warm, moist air from the ocean surface begins to rise rapidly, where it encounters cooler air that causes the warm water vapour to condense and form storm clouds and drops of rain. The condensation also releases latent heat, which warms the cool air above, causing it to rise and make way for more warm humid air from the ocean below. As this cycle continues, more warm moist air is drawn into the developing storm and more heat is transferred from the surface of the ocean to the atmosphere. This continuing heat exchange creates a wind pattern that spirals around a relatively calm centre, or eye, like water swirling down a drain. In 2005, a strong hurricane known as Katrina, reached the coastal areas of America. It was recorded as the sixth strongest hurricane in the history of the United States. At least 1,836 people died in the actual hurricane and the subsequent floods, and total property damage was estimated at $81 billion (2005 USD) Eventually 80% of the city and large tracts of neighboring parishes were flooded, and the floodwaters lingered for weeks. However, the worst property damage occurred in coastal areas of Mississippi beachfront towns. Scientists believe that the increase in severe hurricanes over the past decade is due to natural salinity and temperature changes deep in the Atlantic—part of a natural environmental cycle that shifts back and forth every 40-60 years. Although, some scientists believe that human activities have contributed significantly to global warming, which is increasing air and water temperatures worldwide and making it easier for hurricanes to form and gain destructive force.

2.8.2 Great Eastern Japan Earthquake (Tsunami)

On 11 March, 2011, at 14:46 JST on Friday, the **Great Shake on East Japan** occurred, which was of 9.0 magnitude mega thrust earthquake off the coast of Japan. The epicentre was approximately 72 km east of the Oshika Peninsula of Tôhoku, with the hypocentre at an underwater depth of approximately 32 km. On 1 April, 2011, the Japanese government named the disaster resulting from the earthquake and tsunami the "Great Eastern Japan Earthquake". The earthquake triggered extremely destructive tsunami waves (37.9 metres) struck the Japan after the quake. These waves travelled up to 10 km areas and many small waves reached many other countries after several hours. The earthquake moved Honshu 2.4 m (7.9 ft) east and shifted the earth on its axis by almost 10 cm (3.9 in). The Japanese National Agencies confirmed 13,858 deaths, 4,916 injured and 14,175 people missing, as well as over 125,000 buildings damaged or destroyed. The earthquake and tsunami caused extensive and severe structural damage in Japan, including heavy damage to roads and railways as well as fires in many areas, and a dam collapse. Around 4.4 million households in northeastern Japan were left without electricity and 1.5 million without water. Many electrical generators were taken down, and at least three nuclear reactors suffered explosions due to hydrogen gas that had built up within their outer containment buildings after the cooling system failure.

2.8.3 El Niño and La Niña

El Niño can be understood as prolonged differences in Pacific Ocean surface temperatures when compared with the average value, or in other words, it can be said that the warming or cooling of at least 0.5°C (0.9°F) averaged over the east-central tropical Pacific Ocean. Generally, this anomaly happens at irregular intervals of 2–7 years. The El Niño-Southern Oscillation is often called just "El Niño". El Niño is Spanish for "the boy" and refers to the Christ child, because periodic warming in the Pacific near South America is usually noticed around Christmas.

The sea surface temperature across the equatorial Eastern Central Pacific Ocean will be lower than normal by 3–5°C. At this condition the movement of air is known as La Niña. The name La Niña originates from Spanish, meaning "the girl". La Niña, sometimes informally called "anti-El Niño", is the opposite of El Niño, where the latter corresponds instead to a higher sea surface temperature by a deviation of at least 0.5°C, and its effects are often the reverse of those of El Niño.

EXERCISE

QUESTIONS

1. What are the natural sources ? Describe them with a suitable example.
2. Describe the main threats of forest.
3. Explain the benefits and disadvantages of dams.
4. Define flood, its impacts and their mitigation.
5. Write short notes on forest fire and shifting cultivation.
6. What are the main sources of water?
7. What are the impacts of shifting cultivation?

OBJECTIVE TYPE QUESTIONS

1. The major purpose of most of the dams is:
   (a) electricity (b) irrigation
   (c) flood control (d) all

2. The major disadvantages of dams is:
   (a) flood (b) rehabilitation
   (c) environmental problems (d) all

3. How much forest % is good for ecological balance?
   (a) 33% (b) 66%
   (c) 35% (d) 38%

4. The major cause of forest degradation is:
   (a) forest fire (b) shifting cultivation
   (c) flood (d) all

5. Major sources of water are:
   (a) clouds (b) glaciers
   (c) rivers (d) rain

6. Types of rivers:
   (a) three (b) two
   (c) four (d) all

7. Types of water sources:
   (a) groundwater (b) surface water
   (c) both (d) river

8. Rivers types are:
   (a) snow fed rivers (b) rainfall rivers
   (c) both (d) none

9. The major part of surface water is:
   (a) pollute (b) sweet
   (c) salty (d) none of these

10. 97% of water is stored in the:
    (a) oceans (b) glaciers
    (c) rivers (d) all

11. Surface water sources are:
    (a) sea (b) lakes
    (c) rivers (d) all

12. India's forest area is ... of the total geographical area:
(a) 19.27% (b) 19.6%
(c) 19% (d) 21%
13. The forest of the country can be divided into:
(a) 16 types (b) 15 types
(c) 14 types (d) none
14. State world's total quantum of water is:
(a) 1.4 billion cubic kilometer (b) 1.1 billion cubic kilometer
(c) both (d) none
15. Who laid down the water quality parameters and standards?
(a) BSI (b) ZSI
(c) ISI (d) ISD
16. Kauvery water dispute is in between:
(a) India-Pakistan (b) India-China
(c) Uttarakhand-U.P. (d) Karnataka-Tamil Nadu
17. The longest river of India is:
(a) Ganges (b) Godavari
(c) Krishna (d) Sindu
18. Most flood river in the northeast:
(a) Brahmaputra (b) Ganges
(c) Gandak (d) Tista
19. The rivers in northern part of India are:
(a) snow fed (b) rain fed
(c) both (d) none
20. Man-made dam in Uttarakhand:
(a) Tehri dam (b) Vayasi dam
(c) both (d) none
21. Mega hydroproject are harmful due to:
(a) security (b) natural calamities
(c) flood (d) all
22. Maximum oxygen production habitat is:
(a) forest (b) urban
(c) cities (d) rivers
23. Dams are beneficial for:
(a) irrigation (b) drinking water supply
(c) fish farming (d) all
24. The Indian National River is:
(a) Ganges (b) Krishna
(c) Tista (d) all

25. To solve the water problem, a technique is used known as:
    (a) interlinking the rivers (b) construction of dams
    (c) rivers (d) all

ANSWERS

1. (d)	2. (d)	3. (a)	4. (a)	5. (d)	6. (b)	7. (c)	8. (c)	9. (c)	10. (a)
11. (d)	12. (a)	13. (a)	14. (a)	15. (c)	16. (d)	17. (a)	18. (a)	19. (a)	20. (c)
21. (d)	22. (d)	23. (d)	24. (a)	25. (a)					

3

ECOSYSTEM

3.1 INTRODUCTION

The ecosystem is the structural and functional unit of ecology. An ecosystem is a community of organisms in a given area. It refers to the relationship with their environment. The term ecosystem first appeared in 1935, published by the ecologist A.G. Tansley, who defined it as "*resulting from the integration of all living and non-living factors of the environment.*"

The central theme of ecosystem concept is that the living organisms of a community not only interact among themselves but also have functional relationship with their non-living environment. Thus, the functional and structural system of communities and their environment is called an ecosystem.

An ecosystem represents the highest level of ecological integration, which is energy based and this functional unit is capable of energy transformation, accumulation and circulation. Its main function in ecological sense is to emphasize obligatory relationships, interdependence and causal relations. The earth can be considered a giant ecosystem where abiotic and biotic factors are constantly acting upon each other.

3.2 TYPES OF ECOSYSTEM

An ecosystem may be natural or man-made (artificial). Natural ecosystem may be of several types on the basis of habitats. Generally, it can be classified as terrestrial (i.e., forest, grassland) and aquatic ecosystem, aquatic ecosystem can be divided on the basis of water characteristics i.e., freshwater and marine water. Freshwater can be classified as lentic (stagnant water i.e., lake, pond, ditch) and lotic (running water i.e., river, spring, stream). These ecosystems have been shown in Fig. 3.1.

Difference between Artificial and Natural Ecosystem

Natural ecosystem

- Natural ecosystems are developed under natural conditions.
- Natural ecosystems usually have alternate sources of food available in case one fails.
- In natural ecosystems, inorganic nutrients are returned to the soil from which they were taken.
- Natural ecosystems have no distinct boundaries.

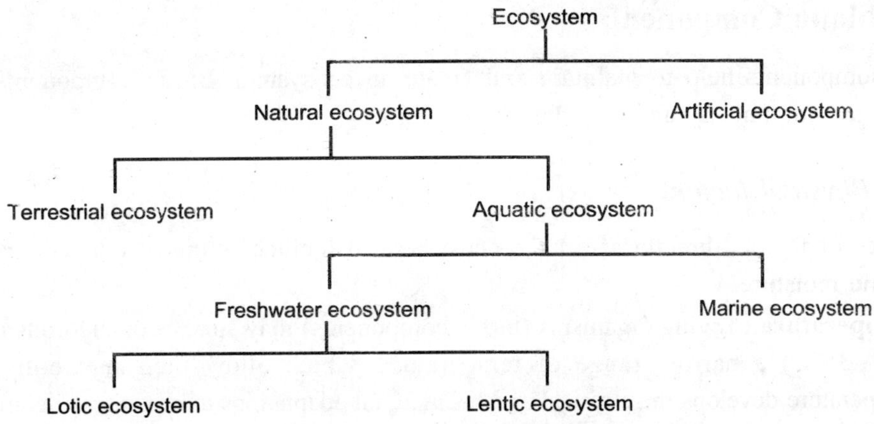

Fig. 3.1: Types of ecosystems.

Artificial ecosystem

- Artificial ecosystems are created and manipulated by human activities.
- Artificial ecosystems usually contain fewer food choices.
- In artificial ecosystems, inorganic nutrients do not return to the soil but are carried away to other places.
- Artificial ecosystems have distinct boundaries.

3.3 COMPONENTS OF AN ECOSYSTEM

An ecosystem comprises two types of components—biotic and abiotic. Biotic components are living beings and include autotrophs, heterotrophs and saprotrophs. Autotrophs are producers (green plants); heterotrophs are consumers (herbivores and carnivores); and saprotrophs are decomposers (microorganisms i.e., bacteria, fungi) Biotic components are connected with each other through food (Fig. 3.2).

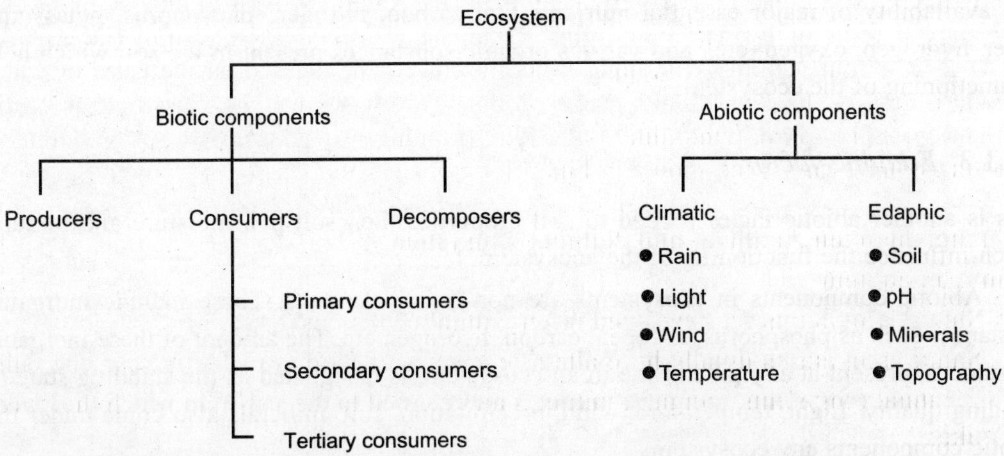

Fig. 3.2: Components of an ecosystem.

3.3.1 Abiotic Components

Abiotic components help to maintain and create an ecosystem. Biotic components linked with abiotic components can be studied following way:

3.3.1.1 *Physical factors*

Such types of factors directly affect the ecosystem, it includes climate, temperature, light, rainfall and moisture.

(a) **Temperature :** Living organisms (biotic components) may survive or vigorous between (20°–35°C) a narrow range of temperature, which allow their metabolisms and temperature develops the physical and behavioural adaptations to organisms i.e. migration to animals in warmer place to avoid extreme cold.

(b) **Water:** Water is an essential requirement of life. The requirement of water varies from organism to organism. The distribution of organism depends upon the extent of the need and special adaptation for conserving water.

(c) **Light:** Light is essential for photosynthesis process in green plants, which sustain the ecosystem through energy flow and food chain. Light also affects a species distribution and colour in nature.

(d) **Humidity:** It is the amount of water vapour present in the atmosphere. Humidity regulates the rate at which water evaporates from the body surface of organisms by transpiration and perspiration.

(e) **Wind:** Wind determines the weather condition. It helps in the dispersal of seeds and fruits of many plants.

3.3.1.2 *Chemical factors*

The availability of major essential nutrients like carbon, nitrogen, phosphorus, potassium, water, hydrogen, oxygen, etc., and various organic substances present in the soil which help in functioning of the ecosystem.

3.3.1.3 *Edaphic factor*

This is another abiotic factor related to soil properties, i.e., soil pH, moisture and texture, which influence the functioning of the ecosystem.

Abiotic components in ecosystems are non-living elements. These include inorganic substances such as phosphorus, nitrogen, carbon, hydrogen etc. The amount of these inorganic substances, present at any given time in an ecosystem, is designated as the standing state or standing quality. Light, temperature, wind, precipitation, soil, minerals also come under the abiotic components any ecosystem.

3.3.2 Biotic Components

Biotic components in any ecosystem are living elements. These can be distinguished on the basis of their nutritional relationships. Biotic components of any ecosystem are given below.

3.3.2.1 *Producers*

Producers are green plants of terrestrial and aquatic ecosystem. They produce their own food in the presence of sunlight. They convert solar energy into chemical energy in the presence of CO_2 and H_2O. Due to this nature of plants, they are also known as autotrophs.

$$6CO_2 + 6H_2O \xrightarrow[\text{Chlorophyll}]{\text{Sunlight}} C_6H_{12}O_6 + 6O_2$$

These autotrophs are called photoautotrophs because they utilize light energy. Another type of autotrophs are chemoautotrophs, which use energy generated in an oxidation reduction process. Microorganisms like *Beggiatoa* and sulphur bacteria are examples of chemoautotrophs.

3.3.2.2 *Consumers*

Consumers are heterotrophs in nature because they depend on producers for food. On the basis of this dependency, consumers can be classified as primary, secondary and tertiary consumers.

(a) Primary consumers: All the primary consumers depend on producers for their food. In this way, these primary producers are generally herbivores in feeding nature. These include, i.e., goat, rabbit, rat, etc.

(b) Secondary consumers: Secondary consumers depend on primary consumers for food. These may be herbivores or carnivores due to their feeding nature i.e., fox, cat etc.

(c) Tertiary consumers: These are the large consumers. Tertiary consumers depend on secondary and primary consumers for food. All the tertiary consumers are carnivores, i.e., tiger, lion, hawk etc.

3.3.2.3 *Decompoers*

Decomposers are microorganisms. They decompose the dead organic materials of producers (plants) and consumers (animals) for their food, and release simple organic and inorganic (minerals) substances into the soil which they produce as by products of their metabolism.

These mineral substances are reused by the producers and cyclic exchange between biotic community and the abiotic environment of the ecosystem (Fig. 3.3). Decomposers are also known as *saprotrophs*. Bacteria and fungi belong to this category.

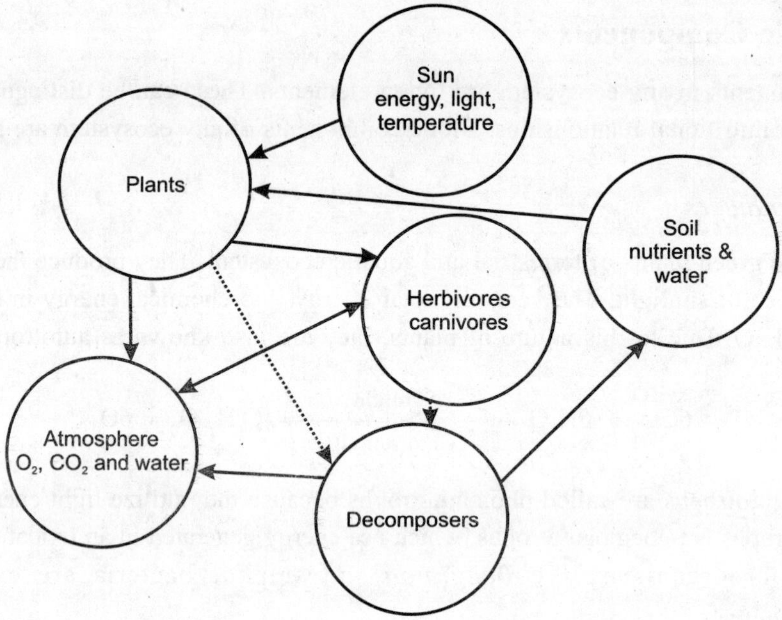

Fig. 3.3: Relationship between the biotic and abotic components of an ecosystem.

3.3.3 Food Chains

The pattern of eating and being eaten from a liner chains called food chain. A food chain always starts with plants and ends with an animal (Fig. 3.4). In a food chain, energy is passed from one trophic level to another. In simple terms, a herbivore feeds on plant and gets energy stored in a new body mass; the rest of the energy is lost as waste or used up. In the same way, when a carnivore eats another animal, only a portion of the energy from the animal food is stored in its tissues.

Generally, food chain may be simple or complex. The simplest food chain involves essentially two links consisting of photosynthetic plants (producers) and animals (consumers) and the decomposers (microbes) that obtain their nutrients from these plants by degrading or decaying them after their death. Complex food chains consist of many links. For example, in lakes, the microscopic plants of the photic zone are primary producers responsible for the production of organic matter by photosynthetic reactions. The plants (producers) and their products may be consumed by certain unicellular or multicelluar herbivorous, zooplankton (consumers), which in turn may be eaten by certain worms and crustaceans (Fig. 3.5). The smaller crustaceans may be devoured by small fish, which are eaten by large fish. The latter in túrn may serve as food for various birds and mammals, whose eventual death and subsequent decomposition by microbes terminates the food chain.

Producers → Primary consumers → Secondary consumers → Tertiary consumers

Fig. 3.4: Food chain in a grassland.

According to Odum, food chain can be of two basic types as follows:

Food chain

Grazing food chain — Detritus food chain

3.3.3.1 *Grazing food chain*

These types of food chains originate from plants and go to grazing animals and then on to animal eaters.

i.e., Grass → Rabbit → Fox → Lion

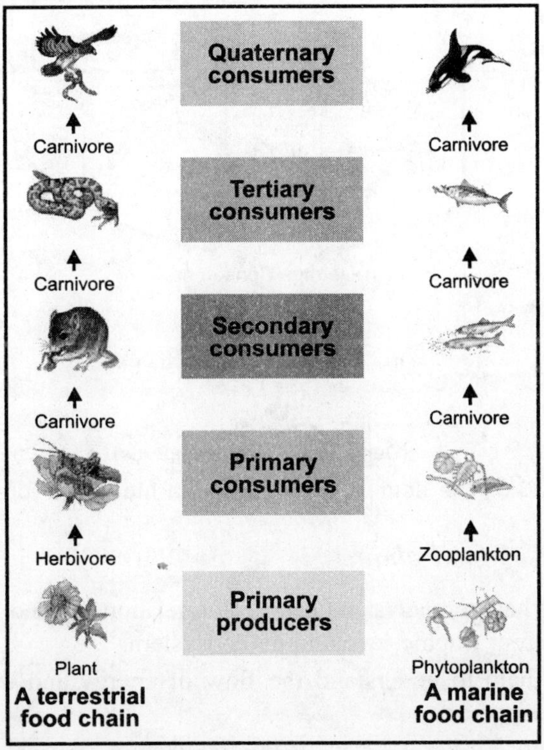

Fig. 3.5: The terrestrial and marine food chain with biotic components of garden and ocean ecosystem.

3.3.2.2 *Detritus food chain*

This type of food chain goes from dead organic matter into microorganisms to organisms feeding on detrivores and their predators (Fig. 3.6).

i.e., Mangrove leaf → Crabs → Molluscs → Nematodes, insects, amphibians → Small carnivores → Top carnivores

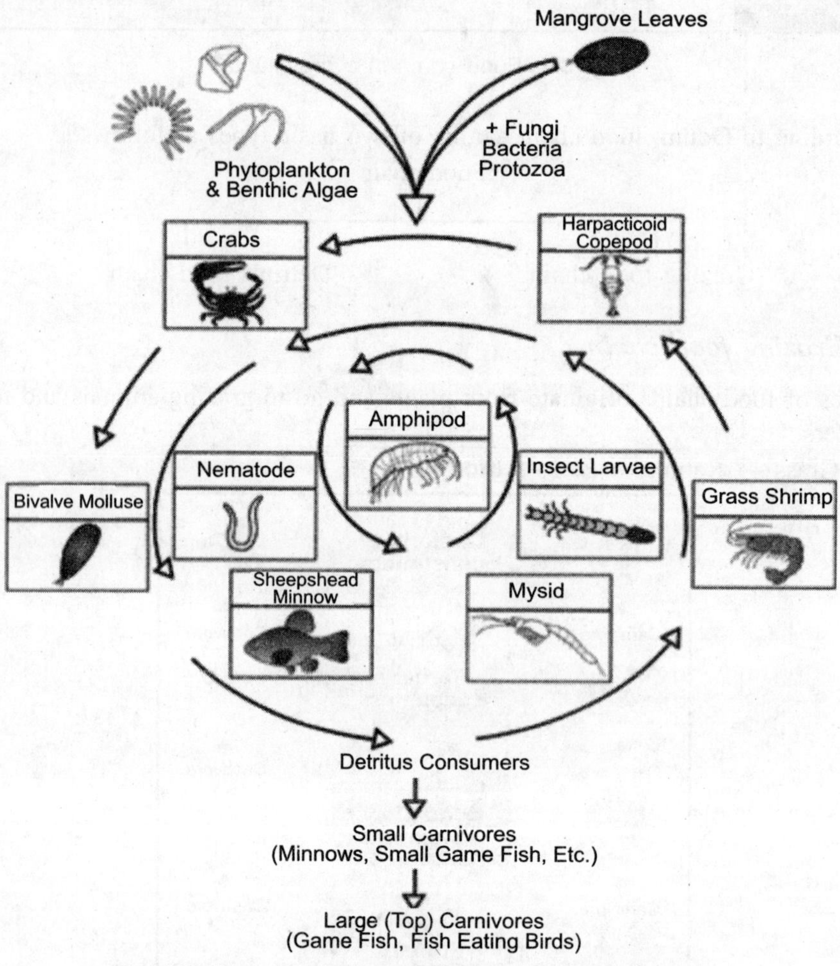

Fig. 3.6: The detritus food chain in a Mangrove forest.

3.3.2.3 *Significance of food chain*

- Food chain studies help to understand the feeding relationship and the interaction between organisms of different trophic levels in an ecosystem.
- Food chains also help to understand the flow of energy and the movement of toxic substances in an ecosystem.

3.3.3 Food Web

There are many food chains surviving in nature or ecosystems. These food chains are interlinked with each other and build a web-like structure. This web is called food web. In other words, we can say that interlinking or matrix of several food chains is called food web. The basic operational principle in a food web is that each species is dependent upon at least some other species, and the numbers of each link species must be sufficient for their continued existence. If these conditions are maintained, the web will exist in an ecological nutritional equilibrium (Fig. 3.7).

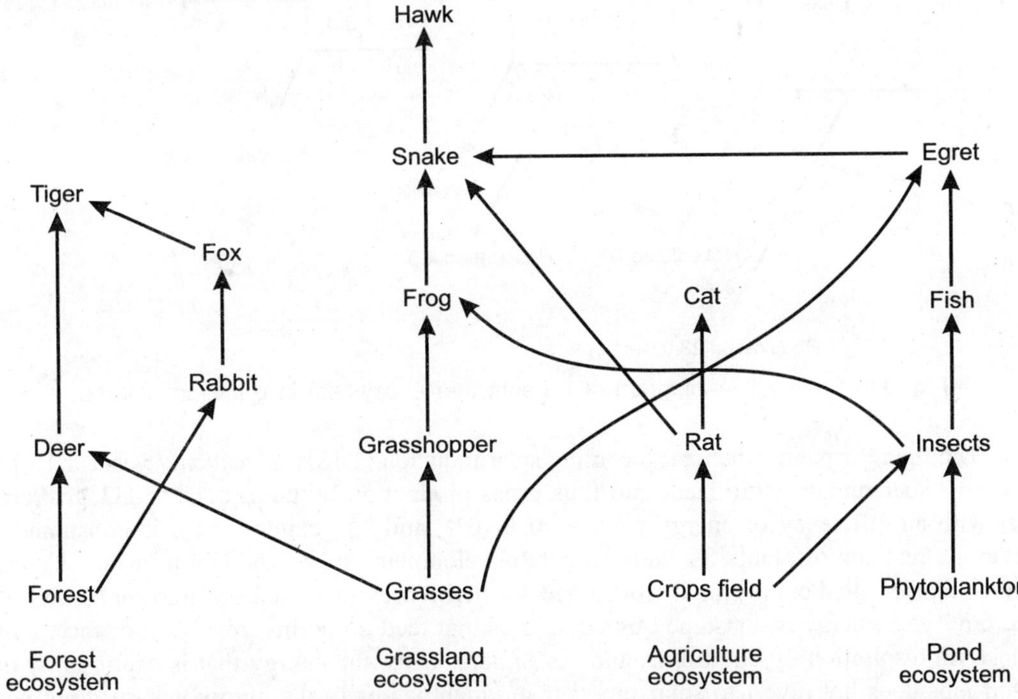

Fig. 3.7: A food web showing interlink among the food chain of different ecosystems.

3.4 ENERGY FLOW IN AN ECOSYSTEM

Energy is the important for life and all the biotic activities cannot regulate without it. The main source of energy on the earth is the sun. Sun produces energy and light, which is absorbed by the green plants. These plants utilize this energy in the photosynthesis reaction and produce the food material in the form of carbohydrate, protein and lipids. These plants are consumed by the consumers, and energy transfers from producers to consumers. Thus, energy transformation in food chain (different trophic levels) is called the flow of energy (Fig. 3.8).

Energy flow in an ecosystem follows two laws of thermodynamics:

1. The first law of thermodynamics states that energy can neither be created nor destroyed, but can only be transformed from one form to another.
2. The second law of thermodynamics states that no energy transformations are 100% efficient.

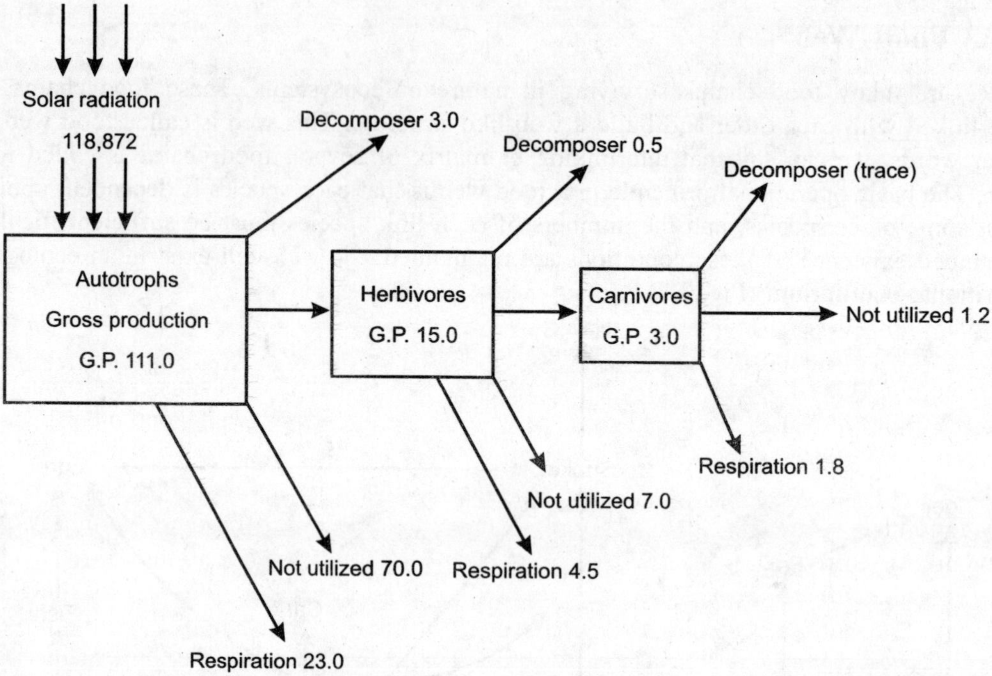

Fig. 3.8: Energy flow diagram of a freshwater ecosystem in g cal./cm^2 /years.

The figure explains, the total incoming solar radiation (1,18,872 gcal/cm^2/ year), 1,18,761 gcal/cm^2/ year remain unutilized, and thus gross production by autotrophs is 111 gcal/cm^2/ year with an efficiency of energy capture of 0.10%, and 23 gcal/cm^2/ year is consumed in metabolic reaction of plants for their growth, development and reproduction. It may be seen further that 15 gcal/cm^2/ year is consumed by herbivores that feed on autotrophs and 3.0 gcal/cm^2/ year energy is consumed by carnivores that feed on herbivores. Most of the energy is lost via respiration by autotrophs and consumers. Thus, the energy that is captured by the autotrophs does not revert to solar input; that which passes to the herbivores does not pass back to the autotrophs.

3.4.1 Ecological Pyramids

A graphical representation of trophic structure or the food chain results into a pyramid-like structure and hence it is called ecological pyramid. In other words, the trophic structure and function at successive trophic levels may be shown graphically by means of ecological pyramids where the producer level constitutes the base of the pyramid and the successive levels, the tiers making the apex.

Ecological pyramids are generally of three types:

1. **Pyramid of numbers**: This represents the number of individual organisms in each trophic level.
2. **Pyramid of biomass**: This indicates the total dry weight and other measures of total amount of living matter in each trophic level.

3. **Pyramid of energy flow**: This shows the rate of energy flow and productivity in different trophic levels.

3.4.2 Pyramid of Numbers

They show the relationship among producers, herbivores and carnivores at successive trophic level (per unit area) in terms of their numbers. On the basis of different ecosystems, pyramid of numbers becomes upright and inverted.

Upright pyramid of numbers, example, the grassland ecosystem shows the upright pyramid. In the grassland, the producers, which are mainly grasses, are always maximum in number. The primary consumers (herbivores) like rabbit, mice, rat, grasshopper etc. are lesser in number than the grasses and the secondary consumers (carnivores) like snake, frog, lizards are lesser than the primary consumers. Finally, the top consumers like hawk, eagle, and other birds are least in number. Thus, grassland ecosystem shows **upright** pyramid in terms of numbers. Similarly, a pond ecosystem shows upright pyramid in terms of numbers (Fig. 3.9(a)).

Inverted pyramid of numbers, example, the forest ecosystem, shows the upright pyramid of numbers. However, in a parasite food chain in forest ecosystem, it shows the inverted pyramid of numbers. In a forest ecosystem, producer (a tree), the primary consumer like birds are maximum in numbers than the trees; the secondary consumers like external parasites (mites, lice and bugs) are maximum in numbers than the birds, and the hyperparasites on lice and bugs are maximum in numbers. Thus, a parasite food chain in the forest ecosystem indicates the inverted pyramid of numbers (Fig. 3.9(b)).

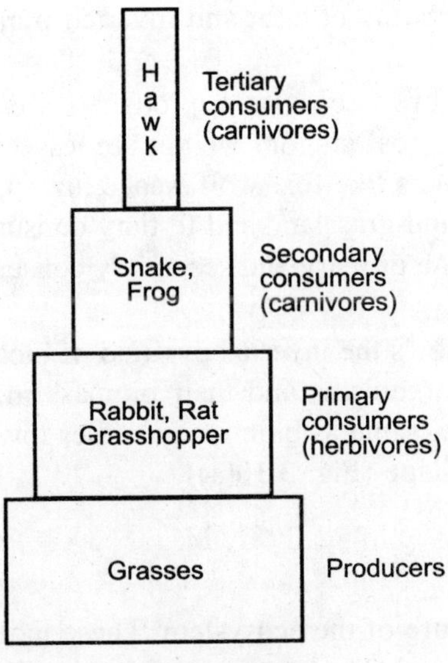

Fig. 3.9(a)

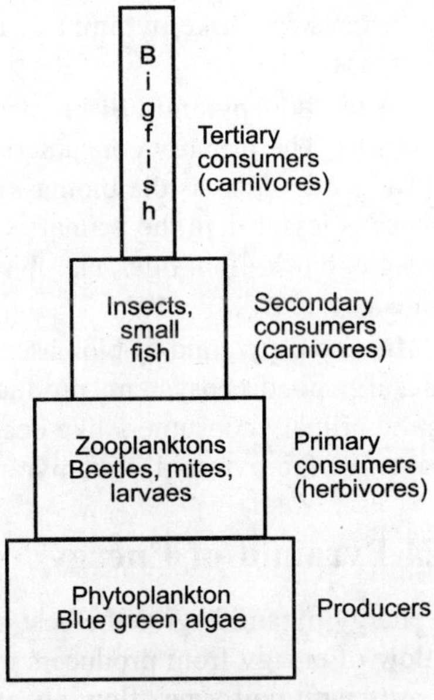

Fig. 3.9(b)

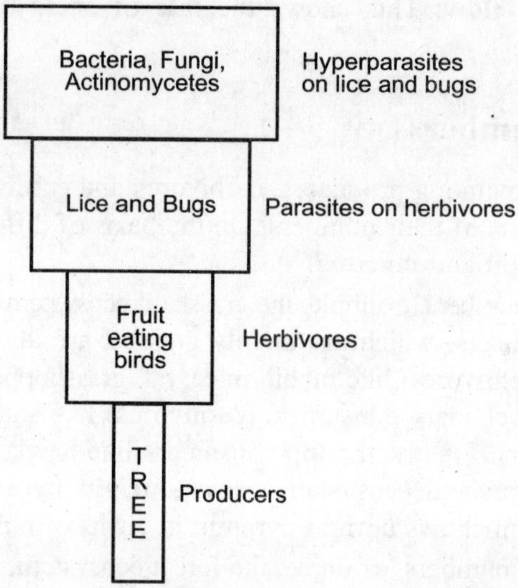

Fig. 3.9(c)

Fig. 3.9 Upright pyramid of numbers (per unit area) **(a)** grassland ecosystem, **(b)** pond ecosystem; Inverted pyramid of numbers (per unit area) **(c)** Parasitic food chain

3.4.3 Pyramid of Biomass

Pyramid of biomass shows the dry weight (biomass) per unit area of different trophic levels of an ecosystem. Like pyramid of numbers, it also shows the upright and inverted pyramid of biomass.

In **upright** pyramid of biomass, in grassland and forest ecosystem, producers are high in biomass. The primary consumers like rabbit, rat, deer, goat etc. dry weight are lesser than the producers, same as the biomass of secondary consumers like fox, wolf, snake, lizard, etc. biomass is lesser than the primary consumers of forest and grassland and tertiary consumers like eagle, hawk, lion, tiger, etc., biomass is lesser than the primary and secondary consumers in an ecosystem.

Inverted pyramid of biomass, a pond ecosystem shows the inverted pyramid of biomass because in pond ecosystem, producers are very small organisms and their biomass in less than the primary consumers like beetle, larves, insects etc. same as biomass increases towards the apex of the pyramid, thus pyramid takes inverted shape (Fig. 3.10(a)).

3.4.4 Pyramid of Energy

The energy pyramids give the best picture of overall nature of the ecosystem. These indicate the flow of energy from producers to consumer trophic level. The pyramids of energy (K cal/ unit area with unit time) flow always show upright, which indicates that energy level from producers to top consumers of any ecosystem always declines.

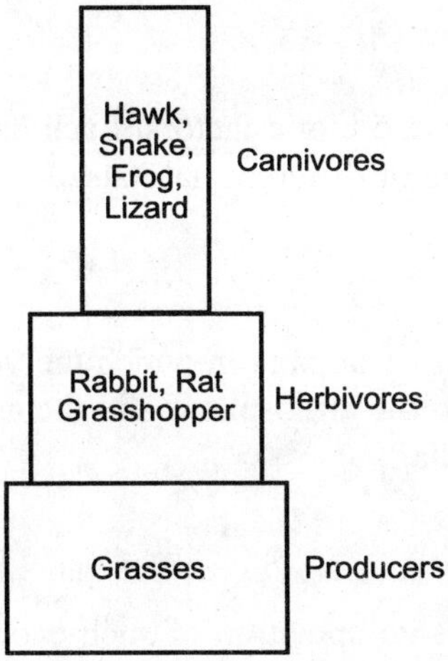

Fig. 3.10 (a)

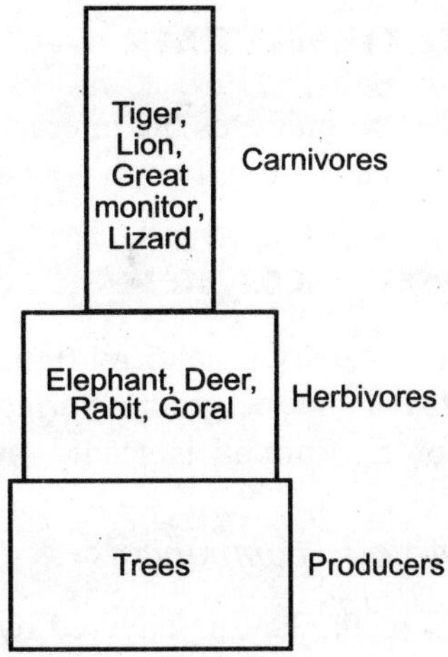

Fig. 3.10 (b)

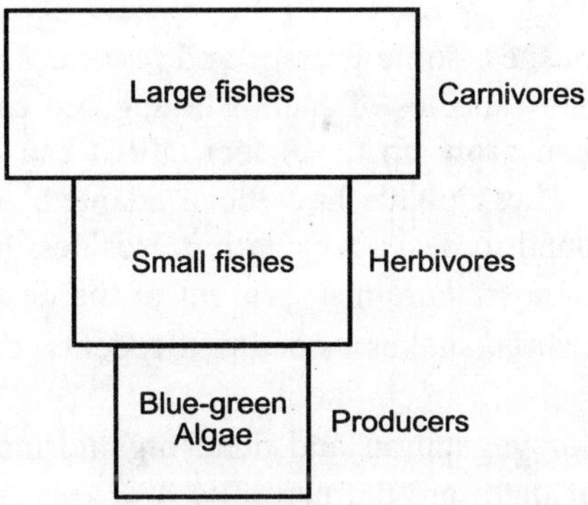

Fig. 3.10 (c)

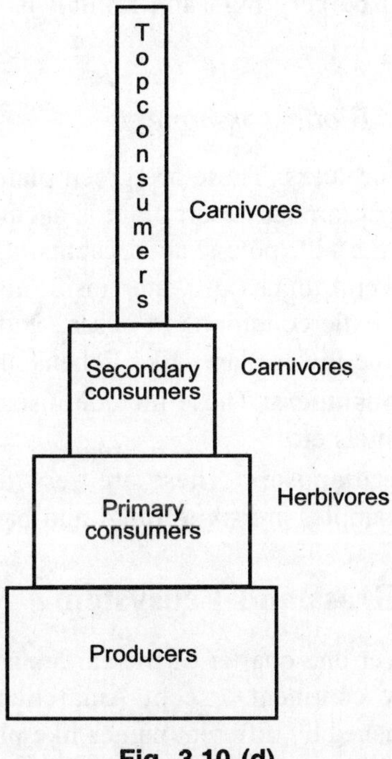

Fig. 3.10 (d)

Fig. 3.10: Upright pyramid of biomass, **(a)** grassland ecosystem, **(b)** forest ecosystem; Inverted pyramid of biomass **(c)** Parasitic food chain and **(d)** Pyramid of energy (k,cal/unit area) in any ecosystem.

3.5 ECOSYSTEMS

Each ecosystem indicates the interaction between abiotic and biotic factors. Each habitat has a different ecosystem. Given below are some ecosystems of different habitats.

3.5.1 Desert Ecosystem

Desert also occupies a major portion of land in India. This is an area in which the vegetation cover is sparse and the ground surface is thus exposed to the atmosphere and the associated physical forces. Rainfall is scanty, infrequent and irregular.

3.5.1.1 *Abiotic components*

About 17% of the land is covered by desert. The species composition of such ecosystem is much more varied and typical due to extremes of both temperature and water factors. The mean max temperature ranges from 39°C to 42°C during summer, whereas the mean min temperature ranges from 25°C to 28°C in winter. In the desert ecosystem, temperature is found to be very high and rainfall is very low. The nutrient cycling is poor due to scanty biota.

3.5.1.2 *Biotic components*

1. **Producers**: These are green plants, specially bushes, some grasses and few trees. The most famous desert plant is cactus; there are many species of cacti. The saguaro cactus is the tall, pole-shaped cactus. The saguaros can grow up to 40 feet tall. It can hold several tonnes of water inside its soft tissues. These plants have been adapted to the climatic conditions of desert, and spines are found in desert vegetation. Besides these, some lower plants like lichens and xerophytes mosses are also present in the desert.
2. **Consumers**: These include insects, reptiles (lizards, snakes), nocturnal rodents, birds, camels etc.
3. **Decomposers**: These are very few due to poor vegetation and dead organic matter. Examples are some fungi and bacteria, most of them are thermophilic.

3.5.2 Grassland Ecosystem

Over one-quarter of the earth's surface is covered by grasslands. Grasslands are found in every continent, except Antarctica. There are several types of grasslands. They are distinguished by different names like plains, prairies, savannas and pampas, where the annual rainfall is between 25-75 cm. The dominant plant species are short and tall grasses and other flowering plants, many of which are perennial with extensively developed roots.

Grassland ecosystem is a type of terrestrial ecosystem. In India the spread of grassland and shrub land is put at 12% of the total landmass. The various components of a grassland ecosystem are as follows:

3.5.2.1 *Abiotic components*

These are the nutrients present in soil and the aerial environment. Thus, elements like carbon, hydrogen, oxygen, nitrogen, phosphorus, sulpher etc. are supplied by carbon dioxide, water, nitrates etc present in air and soil of the area. Some trace elements are also present is the soil.

3.5.2.2 *Biotic components*

These may be categorized as:

(a) **Producers**: These are mainly grasses such, as Cynodon sp. and lemon grasses. Besides them, few herbs and shrubs also contribute to primary production.

(b) **Consumers**:

(i) Primary consumers are herbivores feeding on grasses such as cows, buffaloes, deer, sheep, rabbits, mouse etc. Besides these there as some insects such as *leptocorisa*, *termites*, *millipeds* etc. that feed on the leaves of grasses.

(ii) Secondary consumers are the carnivores that feed on primary consumers (herbivores). These include animals such as snakes, frogs, fox, hyna, lizards and birds.

(c) **Decomposers**: These are microorganisms present in the soil that decompose the organic compounds into simple elements. These include bacteria, fungi (Mucor, Aspergillus, Penicillium, and Rhizopus) and actinomycetes.

3.5.3 Pond Ecosystem

Freshwater ecosystems provide an enormous variety of habitats for aquatic organisms to grow. They may have running water and stagnant waterbody. A pond indeed exhibits a self-sufficient, self-regulating system. If we examine a bottle full of water or a scoop full of bottom mud, it will show the living organisms (plants and animals) and a mixture of inorganic and organic compounds. Thus, the whole system becomes much complex indeed. Various components of a pond ecosystem are as follows.

3.5.3.1 *Abiotic components*

In the pond ecosystem, the chief abiotic components are heat, light, pH value of water and the basic inorganic and organic compounds, such as water itself, carbon dioxide gas, oxygen gas, calcium, nitrogen, phosphate, amino acids, humic acid etc. Some nutrients are present in pond soluble state, but most of them are stored in particulate matter as well as in living organisms.

3.5.3.2 *Biotic components*

Various organisms that constitute the biotic components in the water ecosystem are as follows:

(a) **Producers**: These are autotrophic, green plants. They fix the radiant energy and with the help of minerals manufacture complex organic compounds like carbohydrates, proteins, lipids etc. Producers may be of different types:

(i) Macrophytes: These are mainly rooted large plants, which include partly or complete submerged, floating and emergent hydrophytes. The common plants are the species of Trapa, Typha, Sagittaria, Chara, Hydrilla, Utricularia, Nelumbo etc. Besides these, some free floating plants like Lemna, Wolffia, Azolla, Salvinia etc. are also present in the pond.

(ii) Phytoplanktons: These are micro, floating or suspended lower plants. These include filamentous algae such as *Zygnema*, *Ulothrix*, *Spirogyra*, *Cladophora* and *Oedogonium*. Free floating phytoplanktons such as *Volvox*, *Diatoms*, *Anabaena*, *Eudorina*, *Chlamydomonas* and *Oscillatoria* etc are also present in the pond.

(b) Consumers: These are heterotrophs which depend on producers (green plants) for food. The consumers in pond are distinguished as follows:

(i) Primary consumers: These are herbivores feeding directly on living plants (producers) These are further divided as:

- Benthos: these are herbivores that feed upon the plant remains lying at the bottom of the pond. The benthic population includes fish, insect's larvae, beetles, mites, molluscs, crustaceans etc.
- Zooplanktons: These are chiefly rotifers like *Asplanchna*, *Lecane* etc. although some protozoan such as Euglena, Dileptus etc. and crustaceans like Cyclops, Pilla, Snails etc. are also present in the pond. They feed on phytoplanktons.

(ii) Secondary consumers: These may be herbivores or carnivores which feed on primary consumers. These include fish, insects.

(iii) Tertiary consumers: These are purely carnivores. Large fishes such as game fish are included in this category, which feed on small fishes.

(c) Decomposers: These are micro in size and are known as microconsumers. They decompose the dead organic compounds are (aquatic plants and animals) into simple substances. Thus they play an important role in the return of mineral elements again to the medium of the pond. They are chiefly bacteria and fungi. *Aspergillus*, *Cladosporium*, *Rhizopus*, *Penicillum*, *Alternaria*, *Trichoderma* etc. are most common decomposers in the water ecosystem.

3.6 ECOLOGICAL SUCCESSION

The term "succession" was used by **Hult** in 1885, and he defined it as, "*the natural process by which the same locality becomes successively colonized by different groups*".

Environment always keeps on changing over a period of time due to:

- variation in climatic and physiographic factors; and
- the activities of the species of the communities.

The phenomenon of transition from one biotic community to another is called succession. However, a stage of development is reached during which there appears to be a dynamic balance between all species and their physical environment. This final stage is referred to as the climax ecosystem. Following are some classic examples:

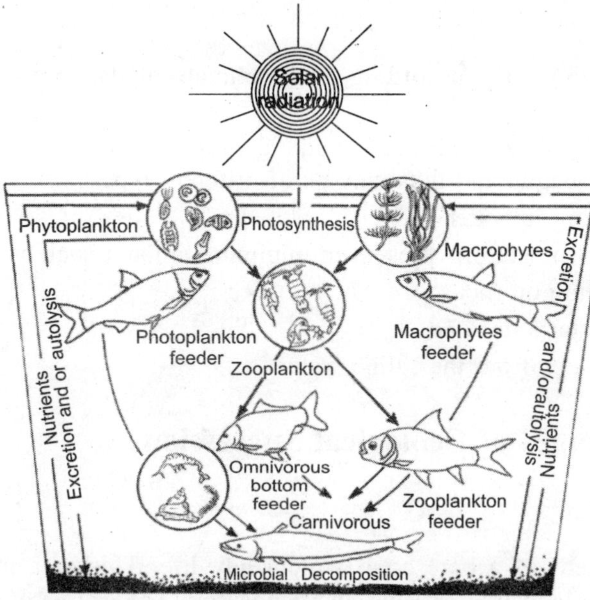

Fig. 3.11: The pond ecosystem.

1. **Primary succession**: If the area has been occupied previously, the process of initial invasion and then the progression from one biotic community to the next is termed "Primary succession". Example is bare rock.
2. **Secondary succession**: When an area has been cleared by fire or humans and then left alone, the surrounding ecosystem may gradually reinvade the area not at once but through a series of distinct stages termed secondary succession.

The major difference between the primary and secondary succession is that secondary succession starts with the pre-existing soil substrate.

3.6.1 Causes of Succession

Causes of ecological succession may be studied in three different steps:

3.6.1.1 *Initial causes*

These are climatic as well as biotic. The former includes factors such as erosion and deposits, winds, fire etc. caused by lightning or volcanic activities. These produce the base areas or destroy the existing populations in the areas.

3.6.1.2 *Continuing causes*

These are the processes as migration, aggregation, competition, reaction etc. These cause successive waves of population as a result of changes, chiefly in the edaphic features of the area.

3.6.1.3 *Stability causes*

Stabilization of the community according to the climate of the area is the chief cause of stabilization.

There are many kinds of succession, which depend upon the environmental conditions, where the process has begun and thus may be of different types:

- Hydrosere: Starting in water body.
- Xerosere: Where moisture is present in minimum amount such as deserts, rocks etc.
- Lithosere: Initiating on rocks.
- Psammosere: It starts on sand.
- Halosere: It initiates from the saline water.

3.6.2 Different Stages of Ecological Succession

3.6.2.1 *Nudation*

This is the development of a bare area without any life. This area may develop due to several causes such as landslide, erosion and deposition. The causes of nudation may be:

- Topographic
- Climatic
- Biotic

3.6.2.2. *Invation*

This is the successful establishment of species in a bare area. This species actually reaches this new site from any other area. This whole process is completed in different stages.

- Migration
- Establishment
- Aggregation

3.6.2.3 *Competition*

After the aggregation of a large number of individuals of a species at a limited place, these develop competition, which may be inter- or intraspecific mainly for space and nutrition.

3.6.2.4 *Reaction*

Reaction is an important stage in succession. The mechanism of modification of environment is through the influence of living organisms. As a result of reactions, changes take place in soil, water, light conditions, temperature etc. of the environment.

3.6.2.5 *Climax* (*Stabilization*)

When the final tar mind community becomes more or less stabilized for a long time and can maintain itself in equilibrium with the climate of the area, this final community is not replaced and is known as climax community and the stage as climax stage.

3.6.3 Hydrosere Ecological Succession

The hydrosere ecological succession is completed in several stages:

3.6.3.1 *Phytoplankton stage*

In the beginning of hydrosere succession, microphytoplankton like blue-green algae, diatoms and bacteria are present in the pond.

3.6.3.2 *Rooted submerged stage*

It is the second stage that comes after phytoplankton stage. A soft mud develops in the bottom of pond, and becomes suitable for the growth of rooted submerged hydrophyte plants like Myriophyllum, Elodea, Hydrilla, Utricullaria, etc.

3.6.3.3 *Rooted floating stage*

After the previous stage, pond bottom becomes shallow and helps the growth of rooted floating plants like Nelumbo, Trapa, Azolla, Lemna, Wolffia etc.

3.6.3.4 *Reed-swamp stage*

This stage is also known as amphibious stage as the plants of community are rooted but most parts of their shoots remain exposed to air. It includes Typha, Sagittaria etc. They have well developed rhizomes.

3.6.3.5 *Sedge-meadow stage*

Due to successive decrease in water level substratum changes and some species, i.e., caser, juncos grow up. As a result of high rate of transpiration, there is much rapid loss of water, and sooner or later the mud is exposed to air as a result of which nutrients like ammonia, sulphides, etc., become oxidized to nitrite and sulphates.

3.6.3.6 *Woodland stage*

By the time of disappearance of marshy vegetation, soil becomes drier for most time of the year. This area is now invaded by terrestrial plants, which are some shrubs (salix, cornus) and trees (poplar, almus).

3.6.3.7 *Forest stage*

The wood land community is rapidly invaded by several trees. In tropical climates with heavy rainfall, they develop tropical rain forest, whereas in temperate regions, these areas develop in mixed forest of Almus, Acer and Quercus. In the region of moderate rainfall, areas develop in tropical deciduous forest or monsoon forest.

Fig. 3.12: The ecological succession.

EXERCISE

QUESTIONS

1. Define ecology and ecosystem. Enumerate different types of ecosystems and explain any two of them in detail.
2. Name and explain various types of ecosystem.
3. Describe the structure and function of a pond ecosystem.
4. Describe the ecological pyramids.
5. What are the ecological components?
6. Describe the ecological succession.
7. Discuss the structure and function of a desert ecosystem.
8. Describe the energy flow in an ecosystem.
9. Define ecology, ecosystem and environment.
10. Differentiate between food chain and food web.
11. Write short notes on the following:
    (i) Food chain

(ii) Food web
(iii) Energy flow
(iv) Detritus food chain
(v) Artificial ecosystem

OBJECTIVE TYPE QUESTIONS

1. Which of the following habitats is most unsuitable for primary productivity?
(a) meadow (b) forested river
(c) cave (d) pond
2. Maximum contribution of O_2 is from:
(a) phytoplankton (b) grassland
(c) herb & shrubs (d) dense forest
3. Ecosystems create:
(a) food chain (b) food web
(c) both a & b (d) none of above
4. In lake ecosystem pyramid of biomass is
(a) upright (b) inverted
(c) anything is possible (d) none of them
5. Man is:
(a) producer (b) primary consumer
(c) top consumer (d) primary & secondary consumer
6. Which one of the following is a top carnivore of grassland food chain?
(a) snake (b) deer
(c) vulture (d) frog
7. The second most important trophic level in lake is:
(a) phytoplankton (b) zooplankton
(c) rhizopus (d) penicillium
8. Which is the most stable ecosystem?
(a) desert (b) ocean
(c) mountain (d) forest
9. Which of the following is the nature's cleaner?
(a) producers (b) consumer
(c) decomposers (d) man
10. Biomass of producers within the specified area will be high in:
(a) forest (b) grassland
(c) lake ecosystem (d) pond ecosystem
11. The succession as a result of changes in communities on the earth's surface is called:
(a) biotic succession (b) climatic succession
(c) geographical succession (d) physiographic succession

12. Pyramid of energy in a pond ecosystem is always:
    (a) inverted (b) upright only
    (c) irregular (d) linear
13. In a food chain, the total amount of living material is depicted by:
    (a) pyramid of energy (b) pyramid of biomass
    (c) pyramid of numbers (d) all of these
14. There are two types of energy flow. These are:
    (a) Z-Y shape (b) X-Y shape
    (c) line -Y shape (d) A-B shape
15. Which one is the correct food chain?
    (a) phytoplankton → zooplankton → fish,
    (b) grass → fish → zooplankton
    (c) zooplankton → phytoplankton → small fish,
    (d) protozoans → algae → zooplankton
16. The term ecosystem was proposed by:
    (a) W.E. Odum (b) A.G. Tansley
    (c) H.T. Odum (d) H.R. Singh
17. The part of the earth and its atmosphere in which organisms live is known as:
    (a) biosphere (b) biomes
    (c) biomass (d) biota
18. Producers in an ecosystem are also known as:
    (a) autotrophs (b) heterotrophs
    (c) chemotrophs (d) none of them
19. India has a forest covered area:
    (a) 10% (b) 15%
    (c) 21.5% (d) 19.4%
20. Xerosere succession is related to:
    (a) land (b) rocks
    (c) pond (d) desert
21. Which one of the following is a biotic component of the ecosystem?
    (a) bacteria (b) plants
    (c) humus (d) fungi
22. Which ecological pyramid is always straight?
    (a) pyramid of energy flow (b) pyramid of biomass
    (c) pyramid of numbers (d) pyramid of biomass & numbers
23. These belong to the category of primary consumers:
    (a) insects & cattle (b) snakes & frogs
    (c) fishes (d) eagles

24. Types of food chain are:
    (a) forest–pond
    (b) grazing-agriculture
    (c) grazing–detritus
    (d) pond–ocean
25. Decomposers are the following groups:
    (a) bacteria & fungi
    (b) algae & zooplankton
    (c) algae & fungi
    (d) none of them
26. Eutrophication is:
    (a) an improved water quality status of lakes
    (b) the result of accumulation of plant nutrients in waterbodies
    (c) a process in the carbon cycle
    (d) a water purification technique
27. The primary producers in a forest ecosystem are
    (a) chlorophyll containing trees and plants
    (b) carnivores
    (c) herbivores
    (d) microorganisms
28. The largest reservoir of nitrogen on our planet is:
    (a) oceans
    (b) biosphere
    (c) atmosphere
    (d) fossil fuel
29. A food web consists of:
    (a) a portion of a food chain
    (b) interlocking food chain
    (c) producers, consumers and decomposers
    (d) a set of similar consumers
30. A trophic level refers to:
    (a) a portion of a food chain
    (b) an organism's position in an ecosystem
    (c) an organism's position in a food chain
    (d) an organism's position in a biome
31. A primary succession on bare rock starts with:
    (a) lichens
    (b) herbs
    (c) trees
    (d) animals
32. A community that starts the process of succession in a habitat is called:
    (a) biotic community
    (b) pioneer community
    (c) abiotic community
    (d) artificial community
33. Which one of the following is an abiotic component of the ecosystem?
    (a) bacteria
    (b) humus
    (c) plants
    (d) fungi
34. Herbivores are consumers:
    (a) autotrophs
    (b) primary consumers
    (c) secondary consumers
    (d) decomposers
35. Which of the following factors influences biotic distributions?
    (a) wind
    (b) light
    (c) temperature
    (d) all of them

36. In terms of a food chain or food pyramid, organisms that trap energy, such as plants are known as:
(a) tertiary consumers
(b) primary consumers
(c) secondary consumers
(d) primary producers

ANSWERS

1. (c)	2. (a)	3. (c)	4. (a)	5. (c)	6. (c)	7. (b)	8. (d)	9. (c)	10. (a)
11. (d)	12. (b)	13. (d)	14. (c)	15. (a)	16. (b)	17. (b)	18. (a)	19. (c)	20. (b)
21. (b)	22. (a)	23. (a)	24. (c)	25. (a)	26. (b)	27. (a)	28. (c)	29. (b)	30. (b)
31. (a)	32. (b)	33. (b)	34. (b)	35. (d)	36. (d)				

4

BIODIVERSITY

4.1 INTRODUCTION

The earth is home to rich and diverse array of living organisms. These living organisms are independent of one another but precisely linked with each other by complex laws of nature. This relationship of living organisms with one another and their physical environment constitutes our planet's biodiversity.

The word biodiversity is a contraction of 'biological diversity'. Biodiversity has been defined by many ecologists and politicians associated with governmental and non-governmental organizations (e.g., Groombridge 2000).

McNeely (1990) considers biological diversity as a concept encompassing all species i.e., plants, animals and microorganisms and the ecosystem and ecological process of which they are a parts. As mentioned in the US Biodiversity Act (1990), the definition of Biodiversity may be "Biological diversity means the full range of variety and variability within and among living organisms, and the ecological complexes in which they occur and encompasse ecosystem or community diversity, species diversity and genetic diversity". Biodiversity is generally considered an 'umbrella term' referring to organisms found within the living world. The term biodiversity came into public vocabulary around 1988 (MaNeely, 1994), and it came into sharp focus only after the convention of Biodiversity (CBD), which was agreed upon by around 157 governments at the Earth submit in Rio de Janeiro, Brazil in June, 1992.

Biodiversity focusses on three levels: ecosystem diversity, species diversity and genetic diversity; all these are interlinked but not synonymous. Maintenance of ecosystem diversity implies maintenance of the species (or at least the most important species), which constitute that ecosystem; however, it is perfectly feasible to maintain species independent of the ecosystems or habitats in which they normally occur. Similarly, maintenance of genetic diversity within a species self evidently implies the maintenance of that species. Species diversity does not distribute uniformly across the globe. Some habitats, particularly in tropical forests among the terrestrial ecosystems—have a great number of species diversity and density than others. Therefore, some countries that are wholly or partially located in the tropical and subtropic regions are known as mega diversity countries (McNeely *et al.* 1990)

Mayer (1992) defined the centre of diversity as merely an area with high species richness, and many centres of diversity have been recognized on global, regional and local scales. On a global scale, four important analyses merit mentioning: (i) Mayers' Hot spots, (ii) IUCN's centre of plant diversity, (iii) Mega diversity centre, (iv) Diversity zones (Barthlott *et al*., 1996). On the basis of species diversity and richness, countries are recognized as mega diversity centre or countries. So far, 12 mega diversity countries have been identified and India is one of them.

This biodiversity is the natural biological capital of the earth and presents important opportunities for all nations. It provides goods and services essential to support human livelihood aspiration and enables society to adapt to changing needs and circumstances. The protection of these assets and their continued exploration through science and technology offer the only means by which the nation of the world can hope to develop sustainability.

4.2 STATUS OF BIODIVERSITY IN THE WORLD

Biodiversity is not distributed evenly on the earth. It is consistently richer in the tropics and other localized regions such as the California Floristic Province, and on the other hand in the Polar regions we generally find fewer species. Flora and fauna diversity depends on climate, altitude, soils and the presence of other species. Biodiversity rich areas are known as hotspots, and these regions have a high level of endemic species. Most of these hotspots are located in the tropics and most of them are forests. These biodiversity hotspots were first identified by Dr. Norman Myers (1988). Hotspot areas that are extremely rich in species have high endemism and are under constant threat. Our earth is a relatively unexpected planet biologically. However it is not feasible to describe all species currently because of the huge number of species on the earth. In the past many scientists estimated a number of species in major groups (Mayer *et al*., 1979; Mayer, 1988; Barnes, 1989; May, 1990) However, almost nothing is known about species ranges and rates of species turnover geographically, so global estimates are uncertain. And what is more intriguing is that a large fraction of thc spccics will disappear before they have been named or described, because the present rate of inventorying or description of new species is quite slow as regard to the present rate of extinction of species.

A large number of earth species are formally classified as rare or endangered or threatened species. However, most scientists estimate that there are millions more species actually endangered, which have not yet been formally recognized. There are about 30 million insects, 15,210 mammals, reptiles and amphibians; 9225 birds, 21000 fishes, about 4, 80,000 plants and 3 million other invertebrates and microorganisms (Solbrig, 1993).

Current estimates range from 13 million to as high as 14 million species in our globe, while 1.75 million species have been recorded scientifically (Heywood *et al*., 1995) (Table 4.1).

Table 4.1: Species number of major taxonomic groups of organisms in the world.

Group	Described species number	Estimated species number	Described/Estimated species number%
Viruses	400,000	40,000,000	1.0
Bacteria	400,000	100,000,000	0.4
Fungi	7,200,000	150,000,000	5.3
Protozoans	4,000,000	20,000,000	20.0

Contd...

Group	Described species number	Estimated species number	Described/Estimated species number%
Algae	4,000,000	40,000,000	10.0
Nematodes	2,500,000	40,000,000	6.3
Molluscs	7,000,000	20,000,000	35
Crustaceans	4,000,000	15,000,000	26.7
Arachnids	7,500,000	75,000,000	10.0
Insects	95,000,000	800,000,000	11.9
Plants (Embryophytes)	27,000,000	32,000,000	84.4
Vertebrates	4,500,000	5,000,000	90.0
Others	11,500,000	25,000,000	46
Total	**175,000,000**	**1,362,000,000**	**12.8**

Sources: Heywood et al. (1995)

4.3 STATUS OF BIODIVERSITY IN INDIA

India is situated at the trijunction of the three realms namely Afro tropical, Indo Malayan and Paleo-Arctic. This assemblage of three distinct realms is the cause of rich and unique biological diversity. India stands 10th in the world, 4th in Asian in plant diversity and ranks 10th in the number of endemic species of higher vertebrates in the world. India occupies only 2.4% of the world's land area, but its contribution to the world's biodiversity is approximately 8% of the total number of species, which is estimated to be 1.75 million. Of these, 1,26,188 have been described in India (Table 4.2).

Table 4.2: Summary of animal biodiversity threats in India.

	Number of Indian species (% of world total)	% of Indian species evaluated	Species threatened in India as % of those evaluated	Number extinct (% of those evaluated)
Mammals	386 (7%)	59%	41%	4 (1.8%)
Birds	1219 (12%)	Unknown	7%	Unknown
Reptiles	495	73%	46%	Unknown
Amphibians	207 (4%)	79%	54%	Unknown
Fresh water fishes	700	46%	7%	Unknown

Based on (Kumar *et al.*, 2000)

The species include flowering plants (angiosperms), mammals, fish, birds, reptiles and amphibians and constitute 17.3% of the total, whereas nearly 60% of Indian biowealth is contributed by fungi and insects (Khoshoo, 1996) At present India has 5 world heritage sites (Kaziranga NP, Keoladeo Ghana NP, Manas Wildlife Sanctuary, Nanda Devi Biosphere Reserve, Sundarban NP), 14 biosphere reserves (Augustamalai and Ankleshwar are new added) and 6 Ramsar wetlands. Amongst the protected areas India has 88 national parks and 490 sanctuaries covering an area of 1.53 lakh km^2. (MoEF 2000) There are 10 biogeographical zones in India (Fig. 4.1). This unique and endemic biodiversity makes India a hotspot of the world. In these biogeographical zones, Himalayan region is rich in endemic species. The Himalayan ranges are about 2500 km long, extending from 73° East to 97° East longitude and from 27° North to 37° North latitude, covering an area of approximately 2,36,000 km^2. These ranges are pull out from few hundred to over 8000 m asl. It is known as the outer Himalaya, middle Himalaya and the inner or great Himalaya ranges. Thus, there are three factors (longitude, latitude and altitude), which provide various microclimate and ecological niches for plants and animals (Khoshoo, 1987). The Western Himalaya extends from Jammu-Kashmir, Himachal Pradesh, and Uttarakhand up to the western border of Nepal; the forest cover of this region is from 4.9 % (J&K) to 63.13% (Uttarakhand) (Anonymous, 1991).

In the Western Himalayan region, where the present study was conducted, Kumaun and Garhwal hills are covered with 65% of the forest land. Forest habitats of this region provide good shelter for this faunal wealth.

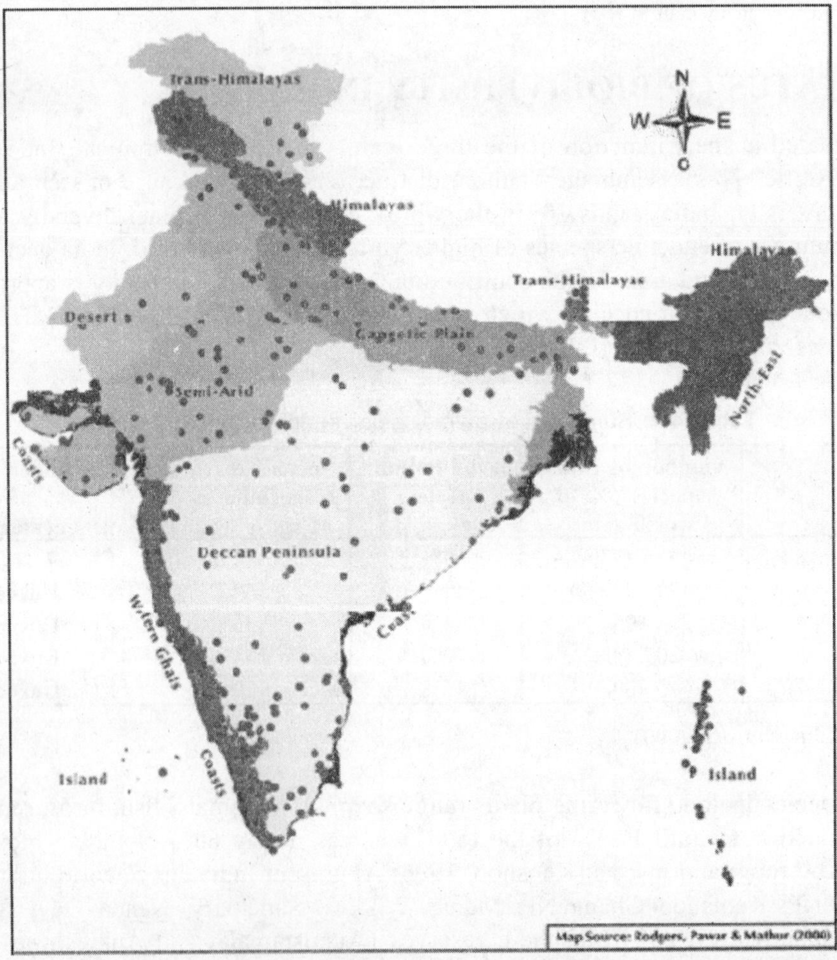

Fig. 4.1: Ten biogeographical zones of India: Trans Himalaya, Himalaya, North East, Gangatic Plain, Desert, Semi-arid zones, Deccan Peninsula, Western Ghats, Coasts, Islands

4.4 TYPES OF BIODIVERSITY

Generally, biodiversity on the earth exists in three levels of organization such as ecosystem, species and genetic biodiversity.

4.4.1 Ecosystem Diversity

Ecosystem diversity depends upon the abiotic factors; an ecosystem develops its own community of living organisms, which shows variation among the species as well as ecosystems. Variation among the ecosystems and their species is referred to as ecosystem diversity. India has 10 biogeographical zones, which provide good ecosystem; several floral and faunal species survive in it.

Some examples of ecosystem diversity are shown in Figs. 4.2 (a) and (b).

(a) (b)

Fig. 4.2: Depiction of ecosystem diversity (a) A fresh water ecosystem, (b) A forest ecosystem.

About 4 % of India's land covering 13,80,00 km^2 is used for ecosystem conservation. The forest department of the state governments manages national parks and sanctuaries.

4.4.2 Species Diversity

There are 75,000 species of animals including 50,000 insects, 4,000 mollusca, 2,000 fishes, 140 amphibians, 420 reptiles, 1,232 birds and 340 mammals and a number of other invertebrates. These are becoming endangered through human encroachment of their habitat poaching. India has applied *ex situ* technique for species conservation in which the botanical and zoological parks are used for species conservation.

For example, the following creatures (coral snake, crow, tiger and pine tree) are the species diversity in nature.

4.4.3 Genetic Diversity

In nature, each species varies from bacteria to higher plants and animals. The variation among them is due to their genetic difference, and their study comes under genetic diversity. The amount of genetic variation is the basis of evolution of new species. It has a key role in the maintenance of diversity at species and community levels. For example, following 3 species of bulbul: (a) Himalayan bulbul, (b) red vented bulbul and (c) red whiskered bulbul are genetically different from each other.

(a)

(b)

(c)

4.5 MEASUREMENT OF BIODIVERSITY

The frequently used diversity measures in ecological analysis are:

4.5.1 Alpha Diversity

Alpha diversity is the number of species in a given area, and the geographical distribution patterns that result from α-diversity are relatively well documented for a wide variety of organisms and have led to such generalizations as an increase in diversity with decreasing latitude.

$$H' = -\sum_{i=1}^{s}\left(\frac{n_i}{N}\right)\ln\left(\frac{n_i}{N}\right)$$

where,

n_i is the total number of individuals of the ith of species in the sample.

N is the total number of individuals.

4.5.2 Beta Diversity

Beta-diversity indicates the comparison of species between two or three habitats. Species distribution also depends on the habitats of their study and analysis comes under beta diversity.

$$\beta_w = \frac{S}{\dot{\alpha}-1}$$

where,

S = the total number of species recorded in the system (i.e. gamma diversity)

$\dot{\alpha}$ = the average sample diversity, where each sample is a standard size and diversity is measured as species richness.

4.5.3 Gamma Diversity

The overall diversity at landscape level, including both alpha and beta diversity studies comes under the gamma diversity.

4.6 CATEGORIZATION OF SPECIES

The International Union of Conservation of Nature and Natural Resources (IUCN) maintains "Red Data Book" carrying information on endangered and vulnerable species of plants and animals. The IUCN has developed categories of species of plants and animals depending upon the degree of danger they are subjected to. The criteria for categorization are as follows:

1. The present and past distribution of species.
2. The decline in the number of population of the species in the course of time.
3. Abundance and quality of natural habitat of the species.
4. The biology and potential value of the species.

On this basis, plant and animal species are categorized as:

4.6.1 Endangered Species (E)

These are the species whose numbers have been reduced to a critical level or whose habitats have been so drastically reduced that they are in immediate danger of extinction.

4.6.2 Vulnerable Species (V)

These are the species that are under threat, such that they have been classified as endangered in near future if the same factors continue to operate.

4.6.3 Rare Species (R)

These are the species with small total population size in the world. These are thinly scattered over an extensive range. Rare species are thus not at present endangered or vulnerable, but are at risk.

4.6.4 Threatened Species

This term is used in the context of conservation of the species, which are not in any one of the above three categories.

4.7 VALUE OF BIODIVERSITY

Each species in ecosystem plays an important role. Their diversity is vital for biosphere health, stability and proper functioning. However, biodiversity is a natural source its value rises in terms of its commercial utility (fuel, fodder, medicine), ecological services (climate, flood control, nutrient recycle) and social (recreational, aesthetic, research).

Following are the major aspects of biological diversity that make it so important for us.

4.7.1 Valuable Natural Resources

The biodiversity may have its values in different sense. Some of these may be direct value (i.e., commodity value), which is obtained on harvesting (i.e., using it), while others are indirect values, which are non-commodity value. The direct values of species are obtained on their destruction (i.e., consumption), whereas indirect values are the benefits derived without destroying the natural resources. Man, since ancient age, has been using plants and animals for food, clothing and shelter. Man has domesticated a number of economically important plants and animal species. However, a large number of useful products still come from wild plants and animals. Man has been cultivating about 2500 plant species, and these are used for fuel, fodder and medicinal purposes. These valuable natural resources are used not only for domestic purposes, but also for commercial ones.

4.7.2 Productive Value

Many products have been synthesized based on the natural products from biodiversity. Many tropical plants produce chemicals that deter herbivores. Tropical indigenous people have discovered many of these plants and use them as a poison or medicine. For example, Calabar bean was traditionally used as a poison in West Africa. This plant is used for the production of insecticide, methyl carbamate. This is similar to bacterium *Bacillus thuringiensis*, which produces toxic proteins that kills certain insects but are apparently harmless to humans. The Neem tree in India has been found to be a source of the insecticide azadirachtin as well as fungicides, spermicide potentially valuable in birth control.

The pharmaceutical industry is dependent on natural products. About 121 prescription drugs are derived from higher plants. These include morphine, codeine, quinine, atropine and digitalis.

4.7.3 Social Value

Man is a social element. He lives in small and big groups, combination of such groups forms a society. For leading a cheerful and enchanting life, they carry on various activities which are acceptable to society. Some of these activities are owning garden, developing flower parks, fishing, cattle and animal domesticating, agriculture etc. These activities indicate the social value of biodiversity. Many of the plants and animals are considered holy and sacred in our country like tulsi, peepal, mango, lotus, bael, cow, snake, bull, peacock and owl etc.

4.7.4 Ethical Value

Although the society today has become money-oriented, yet a substantial portion of it honors the moral values. That is why, contrary to many people using species for their monetary benefits, there are many others who try to protect them. This is so on account of the ethical values in society. After all, we the human beings, being the most intelligent species on earth, cannot ignore our duty to keep the continuance of all forms of life on earth.

Similar to religion, the ethical values are in fact the part of our culture. Therefore to preserve these values, conservation of biodiversity is essential. This can be achieved by boycotting the misuse of animals, plants, flowers, birds etc.

4.7.5 Aesthetic Value

The aesthetic values of our nature ecosystems and landscapes contribute to the emotional and spiritual well-being of a highly urbanized population. Natural and wild landscapes are aesthetically pleasing and provide opportunities to get away from human dominated landscapes. They also provide opportunities for recreational activities such as hiking, canoeing, bird watching and nature photography. Ecotourism (enjoyment of aesthetical value) is an example of aesthetic value of biodiversity. According to Confederation of Indian Industries (CII) statistics, travel and tourism supports 5.8% of the total employment and generates 10.8% of the total exports of the country.

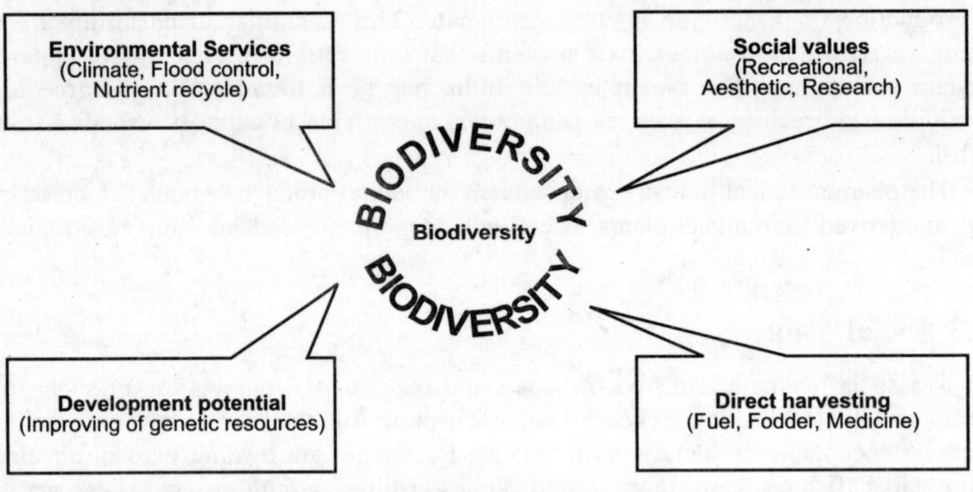

4.8 CONSERVATION OF BIODIVERSITY

It is essential to adopt different approaches for managing our forests and biodiversity. The ecological processes that have maintained the area's biodiversity such as predation, pollination, parasitism and seed dispersal and the herbivores also involve complex interactions between several species of plants and animals.

The corrosion of biodiversity is largely due to habitat loss caused by the expansion of various development projects such as mines, dams, canal construction and urbanization. It is estimated that after independence India lost 4,696 million hectares of forest land to non-forestry purpose (MoEF, 1999). Presently, greenhouse gases (CO_2, CH_4 and CFC) are increasing in the atmosphere, and they have affected global biochemical cycles and the climate system.

As mentioned earlier, India is recognized as a country uniquely rich in all aspects of biodiversity. For any one country in the world, it has perhaps the largest array of environmental conditions by virtue of its tropical location, varied physical features and climatic type. India has the widest variety of biomes. Groves are unique tradition which have been responsible for preserving the pockets of biodiversity in various parts of the country. Such groves occur mainly in the Western Ghats and North Eastern India. The trans Himalayan regions are richest in faunal community and migratory birds in the world due to their geographical conditions and sparse vegetation. The biodiversity of that area is declining in quality and quantity due to the deforestation, grazing and wood cutting.

Generally, there are two basic approaches that are helpful to protect our biodiversity.

4.8.1 *In Situ* Conservation

In this approach, we protect or conserve the species into their own natural habitat/ ecosystems. The emphasis is given to save either the entire protected area or the endangered species. National parks, sanctuaries and biosphere reserves come under the *in situ* conservation techniques. There are 96 national parks and 488 sanctuaries and 15 biosphere reserves in India.

4.8.1.1 *National Parks*

National parks come under the *in situ* conservation techniques. They help to conserve the habitat of endangered species of flora and fauna. India's first national park (*IUCN category II protected area*) was established in 1935 as Hailey National Park, now known as Jim Corbett National Park. In India protected area with national parks is 38,029.18 km^2 which is about 1.16% of total surface of India. As of April 2007, 96 national parks have been established in India. Some of them are Raja ji NP, Nanda Devi NP, Kaziranga NP, Dudwa NP, Ranthambore (Rajiv Gandhi NP), Sariska NP, Kanha NP, Ghana NP, Namidapha NP, Bandipur NP.

4.8.1.2 *Biosphere Reserves*

The programme of biosphere reserve was initiated under the 'Man & Biosphere' (MAB) programme by UNESCO in 1971. The intention of the formation of the biosphere reserve is to conserve the whole ecosystem, so that it could serve as a referral system for monitoring and evaluating changes in natural ecosystems.

The first biosphere reserve of the world was established in 1979. India has established 15 biosphere reserves as shown in Table 4.3.

Table 4.3: List of biosphere reserves in India.

S.No.	Name of Biosphere reserve	Year of Establishment	State
1.	Nilgiri	1986	Karnataka, Kerala & Tamil Nadu
2.	Nanda Devi	1988	Uttarakhand
3.	Nokrek	1988	Meghalaya
4.	Great Nicobar Islands	1989	Andaman & Nicobar Islands
5.	Gulf of Mannar	1989	Tamil Nadu
6.	Manas	1989	Assam
7.	Sunderbans	1989	West Bengal
8.	Simlipal	1994	Orissa
9.	Dibru-Saikhowa	1997	Assam
10.	Dehang-Debang	1998	Arunachal Pradesh
11.	Pachmarhi	1999	Madhya Pradesh
12.	Khangchendzonga	2000	Sikkim
13.	Agasthyamalai	2001	Kerala
14.	Kanchanjunga	2001	Sikkim
15.	Achanakamar-Amarkantak	2005	Madhya Pradesh

4.8.1.3 *Wildlife Sanctuaries*

These are sanctuaries established for the conservation of species. There are 500 wildlife sanctuaries in India covering an area of 117,913.45 km^2, which is 3.59% of the geographical area of the country.

4.8.2 *Ex Situ* Conservation

In this approach, the protection is given to the endangered and rare species in a man-made habitat, other than their natural ecosystems. India has applied *ex situ* technique for species conservation in which the botanical and zoological parks are used for species conservation. About 33 botanical gardens, 275 zoos, deer parks and aquaria have been established.

India has a variety in genetic diversity for prompting conservation using *in situ* and *ex situ* techniques. Several research institutes and genes banks have been established for the conservation of genetic diversity. To conserve the biodiversity, several projects are being successfully implemented in India i.e., project tiger, crocodile and wild ass etc. Several institutes such as wildlife institute (WII), Bombay Natural Historical Society (BNHS) and Salim Ali Centre for Ornithology and Natural History (SACON) are working for the conservation of wildlife.

Fig. 4.3: *Ex situ* conservation.

Updated hotspots are: Tropical Andes, Tumbes-Chocó-Magdalena, Atlantic Forest, Cerrado, Chilean Winter Rainfall-Valdivian Forests, Mesoamerica, Caribbean Islands, California Floristic Province, Guinean Forests of West Africa, Cape Floristic Region, Succulent Karoo, Madagascar and the Indian Ocean Islands, Mediterranean Basin, Caucasus, Western Ghats and Sri Lanka, Mountains of Southwest China, Sunderland, Wallacea, Philippines, Southwest Australia, New Zealand, New Caledonia, Polynesia-Micronesia, Madrean Pine-Oak Woodlands, Maputaland-Pondoland-Albany, Coastal Forests of Eastern Africa, Eastern Afromontane, Eastern Arc Mountains and Southern Rift, Albertine Rift, Ethiopian Highlands, Horn of Africa, Irano-Anatolian, Mountains of Central Asia, Himalaya, Indo-Burma, Japan, East Melanesian Islands, Taiwan, Queensland Wet Tropics (Total: 39) (*source*: www.biodiversity.com)

A brief introduction of Indian Hotspot areas biodiversity is as follows:

Eastern Himalayas: In the whole of Eastern Himalayas (Nepal, Bhutan and China), there are an estimated 9000 plant species, with 3500 (i.e., 39%) of them being endemic. In Indian sector of this area (North-East Himalaya), there are 5800 plant species, about 2000 (36%) of them being endemic.

As regards faunal diversity, 63% of the genera of land mammals in India are known from this area. During the last four decades, two new mammals have been discovered from the region: Golden monkey and Namdapha flying squirrel from Arunachal Pradesh.

North-East area is also a rich centre of avian diversity. More than 60% of the Indian birds are recorded in the North-East. It has two endemic genera of lizards and 35 endemic reptilian species.

Western Ghats: The Western Ghats region is one of the most important biogeographic zones of India, as it is one of the richest centres of endemism. The region has 490 arborescent taxa of which as many as 308 are endemic.

About 1500 endemic species of dicotyledonous plants are reported from the Western Ghats. As regards the fauna, as many as 315 species of vertebrates belonging to 22 genera are endemic. These include 12 species of mammals, 13 species of birds, 89 species of reptiles and 104 species of fish.

EXERCISE

QUESTIONS

1. Define biodiversity. Enumerate its different types.
2. What do you mean by the value of biodiversity?
3. What are the conservation techniques of biodiversity and how can we conserve it?
4. How will you measure the biodiversity? In what respect do alpha, beta and gamma scales differ from each other?
5. Describe the uses and importance of biodiversity.
6. Discuss the structure and function of a desert ecosystem.
7. Explain the causes of biodiversity loss.
8. Explain the patterns of local, national and global level of biodiversity.
9. What are the hot spot areas?
10. Write short notes on:
    (i) Habitat loss
    (ii) Species diversity
    (iii) Endemic
    (iv) National parks
    (v) *In situ* and *ex situ* conservation

OBJECTIVE TYPE QUESTIONS

1. Red Data book refers to:
   (a) a book with red cover (b) a book with a list of extinct species
   (c) a book of a list of rare species (d) book written by Karl Marx
2. Total biogeographical zones in India are:
   (a) 5 (b) 7
   (c) 12 (d) 10
3. Hotspot area refers to:
   (a) hot climate area (b) endemic species area
   (c) endangered species area (d) rare species area
4. Who is the Union Minister of Environment and Forestry?
   (a) Mr. Jai Ram (b) Dr. Manmohan Singh
   (c) Mr.T.R Balu (d) Mr. T.R. Sankarraman
5. Who is the Environment and Forest Minister of Uttarakhand?
   (a) Mr. Bansidhar Bhagat (b) Mr. Bisan S Chaupal
   (c) Dr. Ramesh Pokhriyal (d) Mr. Madan Kausik
6. CITES means:
   (a) Conservation for Indian Tiger and Ecosystems
   (b) Conservation on International Trade in Endangered Species
   (c) Convention on International Trade in Endangered Species
   (d) None
7. Wild Life Protection Act was passed in:
   (a) 1971 (b) 1972
   (c) 1978 (d) 1974
8. Project Tiger was launched in:
   (a) 1965 (b) 1973
   (c) 1972 (d) 1975
9. World Biodiversity day is celebrated on:
   (a) 12 May (b) 23 May
   (c) 24 May (d) 22 May
10. Total protected area in India is:
    (a) 112 (b) 581
    (c) 584 (d) 589
11. Biodiversity Act was passed in India in:
    (a) 2000 (b) 2001
    (c) 2002 (d) still waiting
12. Which among the following species has become extinct?
    (a) Dodo (b) Red headed vulture
    (c) Eagle (d) Hawk

13. IUCN refers to:
    (a) International Union for Conservation Nature
    (b) International Union for Conservation Nature and Natural Resources
    (c) International Union for Conserving Nature
14. UNEP refers to:
    (a) United Nations Environmental Project
    (b) United Nations Environment Program
    (c) United Nations Environment Protection
    (d) United Nations Environment Protection Act
15. Which one is not a wildlife sanctuary:
    (a) Motichur Wildlife Sanctuary (b) Binsar Wildlife Sanctuary
    (c) Naina Wildlife Sanctuary (d) Nanda Devi Wildlife Sanctuary
16. How many hotspot areas have been introduced?
    (a) 12 (b) 24
    (c) 34 (d) 39
17. Indian biodiversity stand at which place at global level?
    (a) 1 (b) 12
    (c) 10 (d) 20
18. National parks come under:
    (a) *in-situ* (b) *ex-situ*
    (c) zoo (d) biosphere reserve
19. Biosphere refers to:
    (a) the wildlife (b) the birds
    (c) the vegetation (d) the ecosystem
20. How many wetlands have been introduced in India?
    (a) 12 (b) 15
    (c) 25 (d) 10
21. Wetland sites are also known as:
    (a) Rem sites (b) Ramsar sites
    (c) Ramsor sites (d) Ramsey sites
22. How many biogeographical zones are there in the world?
    (a) 10 (b) 12
    (c) 14 (d) 20
23. Which is the state flower of Uttarakhand?
    (a) Lotus (b) Burans
    (c) Bram-kamal (d) Rose
24. The term biodiversity was introduced in:
    (a) 1990 (b) 1992
    (c) 1994 (d) 2001

25. How can we conserve our biodiversity?
(a) Through awareness (b) Through indigenous knowledge
(c) Policies and laws (d) Through education

26. Which bird species has become extinct?
(a) Ostrich (b) Dodo
(c) Indian house sparrow (d) Hawks

27. Kaziranga National Park is famous for
(a) tigers (b) elephants
(c) rhinoceros (d) musk deer

28. Which of the following is a cause of loss of biodiversity?
(a) habitat degradation (b) pollution
(c) invasion of non-native species (d) all of the above

29. Spot the site, which is not a biodiversity hotspot?
(a) Brazil (b) California
(c) Central Chili (d) Pakistan

30. Genetic variation between distinct populations of the same species is known as:
(a) species diversity (b) ecosystem diversity
(c) genetic diversity (d) biodiversity

31. Which of the following is an endemic species of India?
(a) Asian elephant (b) whales
(c) panda (d) lion tailed macaque

32. New approach to conservation is the establishment of:
(a) biosphere reserve (b) national park
(c) reserve forest (d) sanctuaries

33. Which group of vertebrates comprises the highest number of endangered species?
(a) birds (b) mammals
(c) reptiles (d) fishes

34. Which of the following is an *in situ* conservation measure taken by India?
(a) Project Tiger (b) Project Elephant
(c) Project Rhino (d) All

35. Which area is rich in biodiversity?
(a) Silent valley (b) NDBR
(c) Ranthambore (d) Datchigam

36. Symbol of WWF is:
(a) Red Panda (b) Giant Panda
(c) Black Panda (d) Teddy Panda

37. Who decides the protected areas?
(a) WCMC (b) UNEP
(c) MoEF (d) BMC

38. Who used the term 'biodiversity crises' for the first time?
    (a) R.H. Whittker
    (b) E.O Wilson
    (c) R.D. Barnes
    (d) N. Mayers
39. Tropical rain forests occur in India in:
    (a) Jammu and Kashmir
    (b) Rajasthan
    (c) Kerala and Assam
    (d) No where
40. Kaziranga National Park is famous for:
    (a) Tigers
    (b) Elephants
    (c) Rhinoceros
    (d) Musk deer

ANSWERS

1. (c)	2. (d)	3. (b)	4. (a)	5. (c)	6. (c)	7. (b)	8. (c)	9. (b)	10. (c)
11. (c)	12. (a)	13. (b)	14. (b)	15. (c)	16. (d)	17. (c)	18. (a)	19. (d)	20. (c)
21. (b)	22. (b)	23. (c)	24. (b)	25. (b)	26. (b)	27. (c)	28. (d)	29. (d)	30. (c)
31. (a)	32. (c)	33. (b)	34. (d)	35. (a)	36. (b)	37. (a)	38. (b)	39. (c)	40. (c)

5

ENVIRONMENTAL POLLUTION AND SOCIAL ISSUES

5.1 INTRODUCTION

In nature, ecological components (abiotic and biotic) are balance between the living organisms and the environment. Disturbance in any component of the environment is likely to have a harmful effect on the ecosystem. Any change in the environment which contributes to its deterioration is called pollution. The term pollution can be defined in various ways; some of the definitions are as follows:

According to Odum, pollution is an undesirable change in the physical, chemical or biological characteristics of air, land and water that will harmfully affect the human life and the desirable species, or may waste or deteriorate our raw material resources.

According to Edwards (1972), "pollution is the release of substances or energy into the environment by man in quantities that can damage either his health or resources".

In other words, pollution is unwanted changes in nature and its components.

5.2 TYPES OF POLLUTANTS

Pollutant can be defined as anything including a chemical, geochemical substance, or a biological product that deteriorates our natural environment.

There are two basic types of pollutants:

- **Non-degradable pollutants**: Such types of pollutants either do not degrade or degrade only extremely slowly in the natural environment. It includes materials and poisons like aluminium cans, mercurial salts, long chain phenolic chemicals and DDT. The non-degradable pollutants also combine with other compounds in the environment to produce additional toxins.
- **Biodegradable pollutants**: These are the pollutants such as the domestic garbage and sewage, which can be rapidly decomposed by the natural processes (by the action of microorganisms such as bacteria and fungi) or take less time for degradation. Heat and thermal pollution can be considered in this category. However, serious problems arise with the degradable types of pollution when the input into the environmental exceeds the decomposition of dispersal capacity of the environment.

POLLUTANTS

Non-degradable Pollutants

Biodegradable Pollutants

5.3 KINDS OF POLLUTION

Pollution can be classified into two types:

(a) **Classification according to pollutants**: There are a large number of pollutants which pollute the environment, and the pollution caused by such substances are sewage pollution, pesticide pollution, smoke pollution, smog pollution, carbon monoxide pollution, chemical pollution, effluent pollution, radioactive pollution, acid rain pollution etc.

(b) **Classification according to environment**: When different parts of the environment are affected by pollution, the problem is generally discussed under water pollution, soil pollution, air pollution.

5.4 WATER POLLUTION

There is about 97% of salt water and 3% freshwater on our planet. Water is one of the most important natural resources and is very essential for the survival of all living organisms. Water pollution can be defined as any unwanted changes or alteration in water quality (pH, DO, acidity, hardness, BOD etc.). These changes may be man-made or due to natural causes. Natural phenomena such as volcanoes, algae blooms, storms, and earthquakes change the water quality and the ecological status of water.

Generally, there are two types of water sources—groundwater and surface water. Sources of surface water pollution are generally grouped into two categories based on their origin (Fig. 5.1).

Point source pollution: It refers to the contaminants that enter a waterway through a discrete conveyance, such as a pipe or ditch. Examples of sources in this category include discharges from a sewage treatment plant, a factory, or a city storm drain etc.

Non-point source pollution: Non-point source (NPS) pollution refers to diffuse contamination that does not originate from a single discrete source. The leaching out of nitrogen compounds from agricultural land, which has been fertilized, is a typical example. Nutrient run-off from the agricultural field or a forest is also an example of NPS pollution.

5.4.1 Groundwater Pollution

The ground level water is about 6.2% of the total water available on the earth. It is about 30 times more than the surface water. There are a number of potential sources of groundwater pollution such as septic tanks, industries and mining etc. Many industries have polluted ground water with arsenic, fluoride and nitrates in different states (Bihar, Rajasthan, and eastern UP) of India.

(a)

(b)

Fig. 5.1: Showing point source of water pollution

5.4.2 Surface Water Pollution

The major sources of surface water pollution are as follows:

5.4.2.1 *Domestic wastes*

It includes food residues, cleaning agents, sewage and detergents. These wastes are rich in organic matter, pathogenic bacteria and other biological pollutants, which decrease the oxygen (DO) and increase the biological oxygen demand (BOD) of the waterbodies like rivers and pollute it.

5.4.2.2 *Industrial wastes*

The major source of water pollution is the waste water discharge from the industries into the waterbodies. Industrial wastes contain toxic chemicals, acids, alkalis, metallic salts, ammonia and radioactive substances etc. which increase the chemical oxygen demand (COD) and pollute the waterbodies. These wastes can be divided into inorganic and organic pollutants. The inorganic pollutants include fine particles of different metals, chlorides, sulphates, cyanides, copper, cadmium, mercury and alkalies. While, organic pollutants include cellulose, carbohydrates, proteins, oils fats, phenols, aromatic compounds etc.

5.4.2.3 *Agriculture wastes*

Agrochemicals like fertilizers and pesticides are a potential source of water pollution. These chemicals increase the production of agriculture fields. Pesticides include insecticides, rodenticides, herbicides etc. There is wide range of chemicals like hydrocarbon, metallic salts, carbonate etc. These chemicals runoff through the rainwater and pollute the waterbodies like rivers, groundwater and contaminate the biotic components of aquatic ecosystem.

For example, DDT compounds run-off through rain water and pollute the waterbody. These compounds are absorbed by the producers and are passed to higher organisms like man through the food chain.

5.4.2.4 *Thermal pollutants*

Thermal pollution of water is caused by the rise in temperature of water. The main sources of thermal pollution are the thermal and nuclear power plants. These release hot water as a waste, which is discharged into the waterbodies. Sudden rise in the temperature of waterbody affects the aquatic ecosystem (biotic factors) and kills fishes and other aquatic animals.

5.4.2.5 *Marine pollution*

Rivers cause marine pollution. Most of the rivers in developing countries are polluted by several human activities, and this water is discharged into the oceans. Petroleum refineries and ship transportation also pollute our oceans through oil spills and waste discharge. War exercises, nuclear test and missiles tests also pollute our marine life.

For example, Gulf War in 1992 polluted our marine life due to the oil spills. "An accidental discharge of petroleum in ocean is called oil spill".

5.4.3 Effects of Water Pollution

The effects of water pollution depend on the types of pollutants present in water. Pollutants change the quality (physical and chemical parameters) of water that makes it unfit for drinking and survival of aquatic life. The main effects of pollutants are mentioned below:

5.4.3.1 *Effects on aquatic ecosystem*

Freshwater is a rich source of oxygen, and this oxygen is present in water in dissolved form, known as dissolved oxygen (DO). It helps in the survival of aquatic ecosystem and organisms. However, due to the pollutants, this dissolved oxygen (water oxygen) reacts with the pollutants and oxidizes them. In this way, freshwater oxygen is utilized for the oxidation of pollutants and the amount of oxygen in the water decreases.

Organic wastes like sewage and garbage which are rich sources of microbes/organic matter, pollute the waterbody. These microbes utilize the water oxygen (DO) in their metabolic reactions. In this way, the biological oxygen demand (BOD) increases and the dissolved oxygen (DO) decreases.

Sometimes the untreated industrial effluents, which are rich sources of hazardous chemicals, pollute the waterbody. These industrial effluents decrease the dissolved oxygen in the water and increase the chemical oxygen demand (COD). This dissolved oxygen utilized by the effluents for the oxidation reaction, known as chemical oxygen demand (COD).

In this way, in the absence of oxygen, water works like as dead body, and without oxygen, producers and consumers of water ecosystem cannot survive.

5.4.3.2 *Eutrophication*

Due to the addition of domestic waste like sewage, phosphate, nitrate etc. and their decomposition in waterbodies, water becomes rich in nutrients. The presence of nutrients in the waterbodies makes it highly productive and stimulates the luxuriant growth of algae in water. The highly productive nature of waterbody is known as eutrophic, and this phenomenon is called eutrophication. The excess growth of algae (algal blooms) utilizes the dissolved oxygen and water becomes unfit for drinking and organism's survival. Reduction of eutrophication should be a key concern when considering future policy, and a sustainable solution for everyone, including farmers and researchers, seems feasible. While, this does pose problems, humans should be aware that natural run-off is common in ecosystems, and should not thus reverse nutrient concentrations beyond normal levels.

5.4.3.3 *Effects on human health*

The presence of pollutants such as heavy metals, pesticides, cyanides and many other organic and inorganic compounds in drinking water or natural waterbodies (rivers, ponds and ground level water) is dangerous. Generally, natural waterbodies are used by the humans after purification. However, due to the addition of substances like pesticides, insecticides, which are used in agricultural field, water becomes polluted. These substances tend to accumulate in the human body and affect our health. Some heavy metals like lead, mercury and cadmium cause various types of diseases. Lead in water may be released from water pipes as lead is used in plumbing. Lead poisoning affects kidneys, reproductive system, liver, brain and central nervous system. It also causes anaemia and mental retardation in children. Mercury dumped into the water is transformed into water-soluble methyl mercury by bacterial action. Methyl mercury accumulates in fish. In 1953, people in Japan suffered from vision, hearing problems and abnormal mental behaviour. This disease is called Minamata disease. Nitrate ions present in the water are harmful to health. Nitrate ions, from nitrogen fertilizers, seep into the waterbodies, from where they may bioaccumulate in the bodies of the consumers and are responsible for Blue baby syndrome and stomach cancer. Fluoride pollution causes defects in teeth and bones—a disease called fluorosis.

5.4.3.4 *Siltation*

Silt is fine sediments carried to rivers by a number of ways including surface run-off. Silt increases the turbidity of water. Although high turbidity from soil particles may be lethal to fishes, turbid waters may affect the productivity of an aquatic environment and the growth of fishes.

5.4.3.5 *Detergents and fertilizers*

The term detergent is applied to washing materials. Detergents are composed of complex phosphates, which eventually break down into phosphates usable by aquatic plants. Although detergents are not highly toxic to fishes, they do cause damage to gills and remove protective mucous from gills, skin and intestine. These chemicals accumulate into the aquatic invertebrates and pass through the food chain and food web (Fig. 5.2).

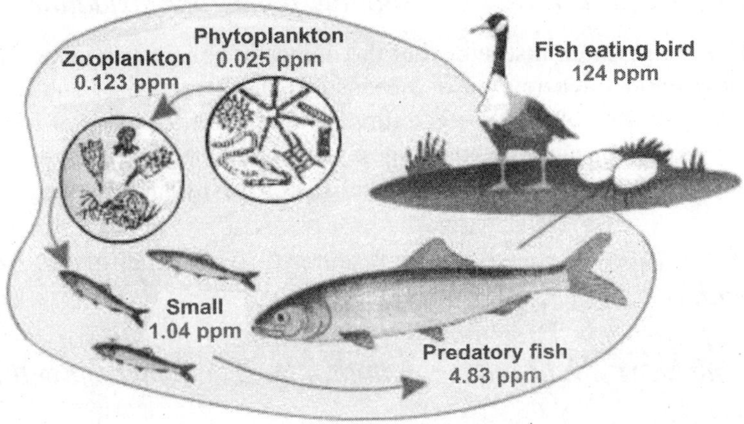

Fig. 5.2: Showing Chemical (DDT) concentration increase in organism along with food chain.

5.4.3.6 *Effect on animals*

Water pollution can occur as a result of natural or unnatural compounds being added to a water source. These can sway the balance of an ecosystem resulting in danger to natural wildlife and aquatic environment. Some animals and plants are highly susceptible to the effects of water pollution:

- Amphibians tend to be very pollution sensitive due to the fact that they absorb chemicals in the water through their skins. This is one reason why many amphibian populations are in distress today.
- With the addition of detergents to the ecosystem, insects that normally have the ability to walk on water will be unable to so.

Younger animals also have a greater sensitivity to chemical compounds, especially those that mimic organic compounds, since they are still undergoing the physiological changes associated with maturation. These effects can range from physiological effects such as suffocation and thinning of eggshells to the alteration of neurochemistry.

5.4.4 Control of Water Pollution

It is easy to reduce water pollution from point sources by legislation. However, due to the absence of defined strategies, it becomes difficult to prevent water pollution from non-point sources. The following points may help to reduce water pollution.

5.4.4.1 *Waste water and sewage treatment*

All waste water and municipal sewage, before dumping into rivers must be treated by primary and secondary methods of treatments.

5.4.4.2 *Treatment of waste water in shallow ponds and oxidation ditches*

It is suggested by the research institutes that the domestic or industrial waste may be stored in a large shallow pond for a few days. Because of the presence of sunlight and organic nutrients in waste, a healthy bloom of algae flourish along with colonies of bacteria. Through this natural process, the bacteria rapidly digest the organic waste, and render it harmless. The effluent may then be used for the irrigation of farms without any danger of pollution.

5.4.4.3 *Adopting integrated pest management to reduce greater reliance on pesticides.*

5.4.4.4 *Uses of nitrogen fixing plants to supplement the use of fertilizers.*

5.4.4.5 *Planting trees would reduce pollution by sediments and will also prevent soil erosion.*

5.4.5 Improvement of Water Quality

Water pollution cannot be altogether checked, but measures can be taken to reduce it.

5.4.5.1 *Primary treatment*

In this method, mechanical screening and sedimentation of undissolved solids in raw sewage is done. But it fails to remove any dissolved substance from water.

5.4.5.2 *Secondary treatment*

Sewage treated in primary treatment is brought in contact with oxygen and aerobic micro-organisms. They break down the organic matter into harmless materials such as CO_2 and H_2O. Further, chlorination is done to reduce the bacteria. It may be further treated to tertiary level. Zero measures are adopted for secondary treatment. These are: (i) trickling filter method and (ii) activated sludge process.

(i) **Trickling filter method**: In this case, sewage water passes through a thick bed of gravel stones, so that bacteria consume most of the organic matter.

(ii) **Activated sludge process**: Here the sewage water is pumped into an aeration tank. The tank contains sludge consisting of bacteria and algae. The bacteria are able to decompose most of the organic matter and algae produces oxygen to promote the growth of these decomposers.

5.4.5.3 *Advanced water treatment or temporary treatment*

Finally, in the third step, salts like nitrates and phosphates are removed by precipitation technique.

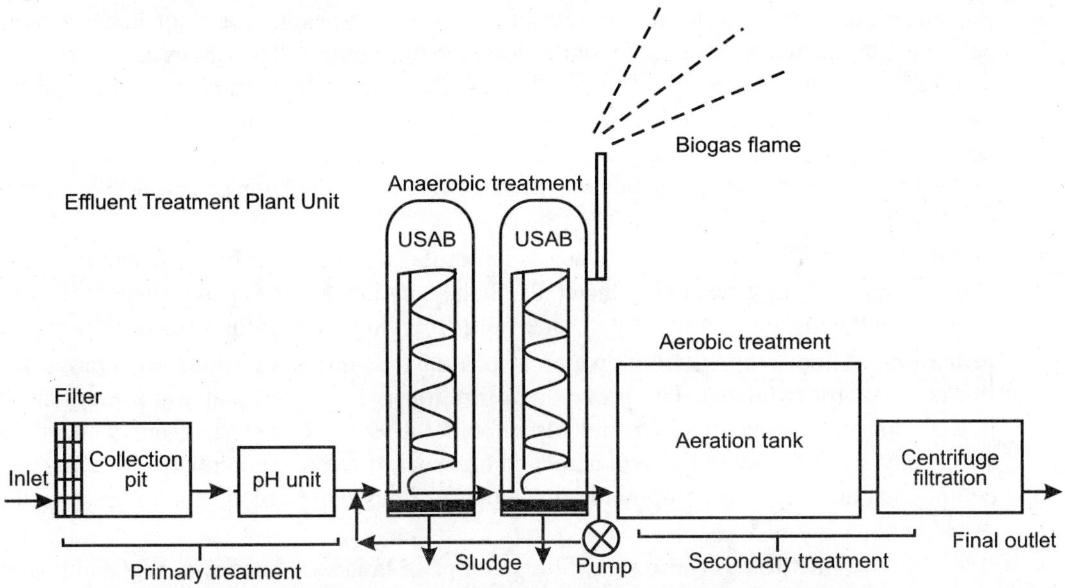

Fig. 5.3: The simple unit of Effluent Treatment Plant (ETP).

5.4.5.4 *Municipal water purification*

There are four main processes of municipal water purification system, which ensure the supply of non-polluted drinking water. These processes are flocculation, sedimentation, filtration and disinfection (Fig. 5.4).

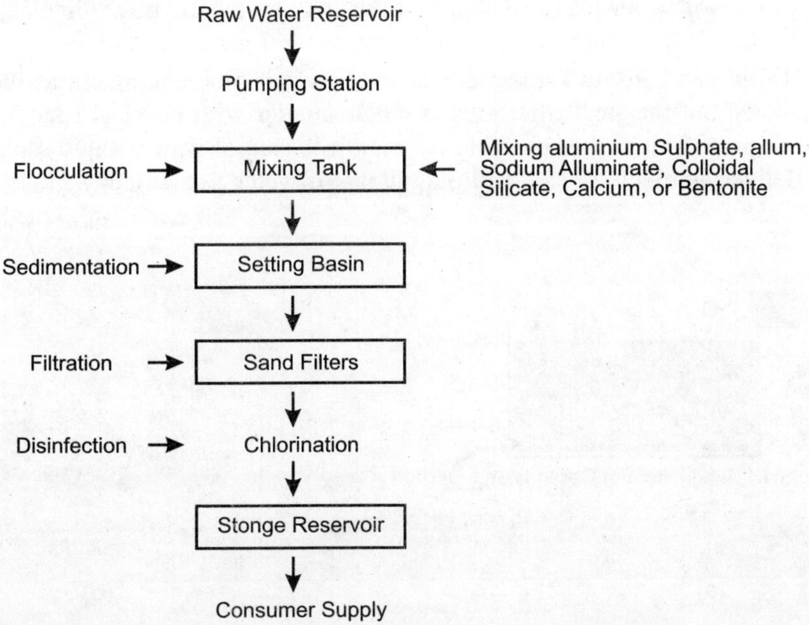

Fig. 5.4: The municipal water purification.

(a) **Flocculation**: Municipal waste water is collected in large tanks for a sufficient time to permit large particulate matter to settle down at the bottom. This material is removed and then the water is treated with flocculants such as aluminium sulphate, alum, sodium aluminate, colloidal silicate calcium, which react with precipitates, making it easy for decomposition by the microorganisms. The surface suspended organic matter settle down to the bottom of the tanks. In this way, most substances that impart turbidity to water get coagulated.

(b) **Sedimentation**: After coagulation, water is left in the settling basin for sufficient period to allow sedimentation of remaining materials. Sedimentation, however, considerably reduces the microbial population of the water aside from removing most of the suspended particles.

(c) **Filtration**: After sedimentation, water is passed through sand filters to remove the flocks of microorganisms. The process of filtration is highly critical and important, as it can remove protozoan cysts and also about 98-99% of bacteria from water. The water may also be filtered through activated charcoal to remove potentially toxic organic compounds and organic compounds that impart undesirable colour and/or taste to the water (Fig. 5.5 a & b).

(i) **Slow sand filters**: These sand filters comprise layers of fine sand, fine and coarse gravel. The floor at the base is made of tiles provided with perforations for intake of filtered water. The uppermost fine sand layer acts as a biological filter, as its interspaces are clogged by colloidal floccules in water, encapsulated microorganisms, and algae, and retains bacteria, making the water free from them. The efficiency of filter is further enhanced as the negatively charged bacteria are held by the positively charged colloidal material in the layer. This filter normally removes 98% of the bacterial population present in water. Nearly 5 to 7 million gallons of water per acre area of filter may be purified by this method, but this method is slow and generally requires several acres of land.

(ii) **Rapid sand filters**: This method involves the same equipment as the slow sand filters, but the sand filter area is much smaller with provision for frequent back washing. This makes possible the rapid filtering of water. Rapid sand filters may deliver as much as 150 million gallons of water per acre/day.

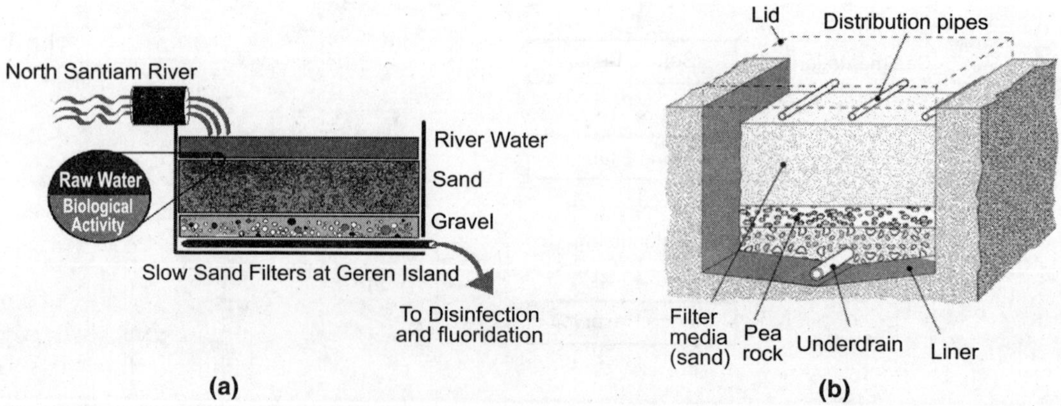

Fig. 5.5.(a) and (b): The sand filters with their components.

(d) Disinfection: Disinfection is the final step of municipal water purification, and it ensures that no pathogenic microorganisms are present in domestic supply of water. In this process, sodium or calcium hypochlorite (NaOCl or $CaOCl_2$ respectively) may be used to disinfect water. However, for larger cities, chlorination (use of chlorine) has been the traditional method of disinfection. Chlorination often removes bacterial contamination, since chlorine reacts with water to produce hypochlorous and hypochloric acids which are potent microbicides. Chlorine is obtained as a liquid under pressure, but released into water as gas. It easily dissolves in water and, in addition to killing the microbes, it is also effective in oxidizing organic matter. However, the greater the organic matter in water, the higher the chlorine demand for a given amount of water, i.e., the chlorine demand for different waters varies depending upon the organic matter of the water.

After disinfection, the water is, finally stored in large reservoirs and is supplied for domestic consumption through gravity taps.

5.4.6 Biochemical Oxygen Demand (BOD)

Biochemical Oxygen Demand (BOD; sometimes called Biological Oxygen Demand), represents the amount of dissolved oxygen needed by the microorganisms in a particular amount of water at 20°C in 5 days. The microorganisms polluted or domestic sewage (organic matter) polluted water is expressed in terms of BOD. The magnitude of BOD is related to the amount of organic material present in the sewage polluted water. A medium or strong sewage (having more oxidisable matter) has high BOD value, and a weak sewage (having less oxidisable matter) has low BOD value. Thus, BOD determinations provide the best information concerning the strength of sewage/ organic waste and the efficiency of sewage treatment processes.

5.4.6.1 *BOD Determination*

The BOD is determined by dilution of a measured amount of sewage with water that has been saturated with oxygen, and incubation of the mixture at 20°C along with a control of dilution water alone. After 5 days, the residual oxygen in both control and test samples is measured by titration. The difference represents the oxygen consuming capacity of the sewage and is calculated to be expressed as parts per million (ppm) of oxygen taken up by the waste. The strength of the sewage in terms of pounds of BOD is calculated as follows:

$$\text{BOD} = (\text{ppm5-day BOD} \times \text{gallons of sewage} \times 8.34)/100{,}000$$

5.4.6.2 *BOD effect in domestic water*

Whenever appreciable amount of strong sewage is emptied into natural waters such as stream, ponds lakes, etc., the 7-8 ppm of free oxygen, which is normally present in water, is quickly utilized to meet the oxygen demand of sewage. This results in a drop in oxygen level in water. When the oxygen drops below 3 ppm, the fish either leave or die. The anaerobic condition is attained and hydrolysis, putrefaction and fermentation by microorganisms follows, with the result that the body of water becomes malodorous and cloudy and hence unsuited for recreational use, unfit for drinking, and other purposes.

5.4.7 Water Pollution: Case Study

5.4.7.1 *Arsenic pollution*

Heavy metals like mercury, arsenic, lead, cadmium and copper etc. are serious toxic water pollutants. West Bengal and Bangladesh are severely contaminated by the toxic heavy metal, arsenic. The first report of arsenic contamination pollution in West Bengal came in 1978 and in Bangladesh in 1993, where it was found to be even more widespread. The accumulation of arsenic indicates white and black spots called melanosis, which started moulting the skin. The spots were later found to get converted into leprosy-like skin lesions encrusting the palms and soles, eventually rotting into gangrenous ulcer. The WHO has prescribed the maximum permissible limits of arsenic as 10 mg/l. The 24 Paraganas, Hoogly and Murshidabad districts as also Behala and south eastern fringes of Kolkata lie in arsenic risk zone.

Excessive use of lead arsenate and copper arsenite as pesticides in high yielding varieties of summer paddy and jute crop seems to be the major cause of arsenic pollution. Now the arsenic contaminated tube wells in the state are being painted red, while safe water tube wells are painted green for use by people.

5.4.7.2 *Mercury pollution*

Mercury enters water naturally as well as through industrial effluents. It is a potent hazardous substance. Both inorganic and organic forms arc highly poisonous. First, mercury pollution was introduced in Japan and Sweden and it was responsible for Minamata disease that caused several deaths. In Sweden, many rivers and lakes are already polluted due to widespread use of mercury compound as algaecides and fungicides in paper and pulp industries, and in agriculture. In India, several cities like Delhi, Kanpur and Agra have located several industries making switches, batteries, thermometer, fluorescent light tubes and high intensity street lamps etc., which use mercury, and their untreated effluent pollute the rivers (Yamuna and Ganga). This mercury compound enters into the waterbody and is metabolically converted into methyl mercury. Methyl mercury is highly persistent and thus accumulates in the food chain.

5.4.7.3 *Fluoride pollution*

Fluoride is also regularly present in water and soil besides air. In Haryana and Punjab, consumption of fluoride rich water from wells caused endemic fluorosis. In Andhra Pradesh also high fluoride content water caused dental fluorosis. On an average, 40 – 50 million Indian are affected with fluorosis. In our country, this problem has become more severe in Rajasthan. Fluorosis is prevalent in districts of Jodhpur, Bhilwara, Jaipur, Bikaner, Udaipur, Nagaur, Barmer and Ajmer. Many people in Rajasthan have humped back due to high fluoride content in water sources and in arid and semiarid zones. Fluoride is not absorbed in the blood stream. It has an affinity with calcium and thus gets accumulated in bones, resulting in the mottling of teeth, pain in bones and joint and outward bending of legs from the knees—knock-knee syndrome.

5.4.7.4 *Yamuna pollution*

The Yamuna river passing through 22 km in Delhi was once described as the lifeline of the city. About 60 million people depended on Yamuna water, but today it has become one of the dirtiest rivers in the country. Yamuna is one of the most polluted rivers in the world, especially around Delhi, which dumps about 57% of its waste into the river. Delhi alone contributes around 3,296 million litres per day of sewage in the river. Several rivers like Tons, in Uttarakhand, Hindon in Uttar Pradesh recharge the river Yamuna till Delhi. The main causes behind Yamuna river pollution is untreated industrial waste and sewage. However, many industries which were polluting Yamuna have been removed or shifted. Now the problem is due to sewage because Delhi generates about 3,267 million litres per day (mld) of sewage, while the city's installed waste water treatment capacity is only 2,330 mld. More than 937 mld of waste that is untreated is allowed to enter the river everyday. According to the CPCB, 70% of the pollution in river is from untreated sewage. The remaining 30% is from industrial source, agricultural run-off, garbage, etc. To solve this problem, Government of India (GOI) started the Yamuna Action Plan in 12 towns of Haryana, 8 towns of Uttar Pradesh, and Delhi, which is being implemented since 1993 by the National River Conservation Directorate (NRCD) of the Ministry of Environment and Forests. The Japan Bank for International Cooperation (JBIC) is participating in the Yamuna Action Plan. The Indian government plans to repair sewage lines to improve the water quality of the river.

Fig. 5.6: Yamuna pollution.

5.4.7.5 *Mumbai oil spill in 2010*

In 2010, Mumbai oil spill occurred after two Panamian-flagged ships, MSC Chitra and MV Khalijia-III collided off the coast of India. MSC Chitra, which was out borne Mumbai on 7.08.2010 at around 9:50 A.M local time from South Mumbai's Nava Sheva port collided with the inbound MV Khalijia-III, which immediately threw out the oil containers from MSC Chitra into the Arabian Sea. Khalijia-III was apparently involved with another mishap on July 18, 2010. It had a cargo of 1,219 containers holding 2662 tonnes of fuel, 283 tonnes of diesel and 88,040 litres of lubricant oil. Thirty-one containers had pesticides in them. The Chitra tilted sharply under the impact of collision, resulting in the oil spill and containers of pesticide bobbing off on the sea.

The Navy and the Coast Guards carried out anti-pollution operations to check and neutralise the oil spill. Six coast guard vessels and a helicopter with anti-pollution dispersal spray systems were pressed into service on 8th August 2010 to contain the oil spill. The formation of a thick layer of poisonous oil in the seawaters inflicted a distressing blow to the bionomical equilibrium by rendering fishes, turtles and other species immobile due to its high viscosity. About four billion dollars were lost by the oil spill pollution within a week. The Mumbai Port lost Rs. 3.5 crore per day, adding up to over Rs. 17 crore. The Federation of Indian Exporters pegged its losses at Rs. 12.5 crore a day with cargo stuck and deliveries delayed. In all, losses of over Rs. 90 crore stemmed before the waterway became fully operational, and that took another month. Around 60 fish markets across Mumbai were empty for one week, and the industry has had an estimated loss of Rs. 60-80 crore. Meanwhile, there is no count of the damage to the environment, the marine life and the many trades that depend on these waters for sustenance.

Fig. 5.7: The impact of oil spill in Mumbai.

5.5 AIR POLLUTION

Air pollution problems are generally associated with the arrival of industrial revolution, population explosion (urbanization), extensive industrial development and automobile revolution.

The release of poisonous gases like carbon monoxide, nitrous oxide and combustion products from industries and automobiles disturb the balance of air components, as well as volcanic eruption, dust storms, forest fire etc, which make the environment detrimental to our health.

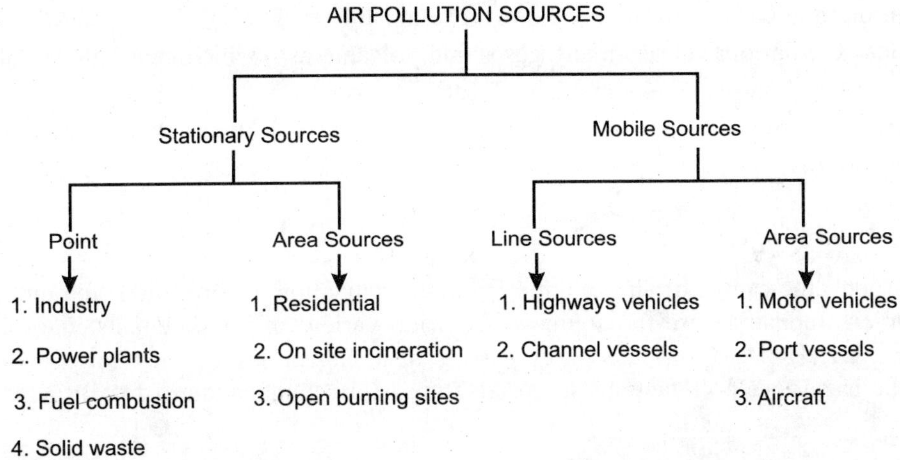

Fig. 5.8: Classification of air pollution sources.

5.5.1 Definition of Air Pollution

(a) Air pollution may be defined as any atmospheric condition in which certain substances are present in such concentrations that they can produce undesirable effects on man and his environment.

(b) Any unwanted changes in air components by the pollutant i.e., dust fumes, gas, mist, odour, smoke is known as air pollution.

Air pollutants can also be of *primary* or *secondary* nature.

Primary air pollutants are the ones that are emitted directly into the atmosphere by the sources (such as power-generating plants).

Secondary air pollutants are the ones that are formed as a result of reactions between primary pollutants and other elements in the atmosphere. Examples of secondary air pollutants are ozone, formaldehyde, PAN (peroxy acetyl nitrate), smog (photochemical or smoke induced), acid mist.

Possibly one of the most important characteristics of air pollutants is their transboundary nature—they can easily travel and affect the areas far away from their points of origination.

5.5.2 Causes of Air Pollution

There are two main causes or sources of air pollution:

(i) Natural

(ii) Anthropogenic

5.5.2.1 *Natural causes or sources*

Following are the natural causes of air pollution:

(i) Natural contamination usually present in the air like pollen, fungal spores, cysts, bacteria and marsh-gas (methane).

(ii) Carbon monoxide from the breakdown of methane.
(iii) Volcanic eruptions release many gases and volcanic ash, which cause air pollution.
(iv) Forest fire releases smoke and harmful traces gases.
(v) Electric storms and solar flares pollute the air by the harmful chemicals they produce.
(vi) Dust storms.

5.5.2.2 *Anthropogenic causes*

The anthropogenic causes of air pollution are industrialization, urbanization, automobile and deforestation. Industries are the sources of a wide variety of air pollutants like SO_2 gas, NH_3, NO_2, HF, HCl gas and H_2S gases besides dust, fumes and tar etc. Automobiles also release the hazardous pollutants by the production of the combustion of fossil fuels.

(a)

(b)

Fig. 5.9 (a) and (b): Stationary sources of air pollution.

5.5.3 Classification of Pollution Sources

The anthropogenic sources of air pollution can be categorized into three types:

(a) **Non-point sources/mobile sources**: The mobiles sources of air pollution are the automobiles.

(b) **Stationary sources**: The stationary sources of air pollution are best exemplified by industries as they add pollutants to the air at particular points from their chimneys.

(c) **Area sources**: Towns and cities add smoke and gases over a wide area and so called area sources of air pollution.

Major air pollutants:

1. **Carbon monoxide (CO)**: Carbon monoxide results due to incomplete combustion of fuels from automobiles or industries. CO is an odourless, colourless, poisonous gas that comes mainly from motor vehicles (2 strokes). Carbon monoxide interferes with the blood's ability to carry O_2 to the brain, heart and other tissues, and it is particularly dangerous to people with existing heart disease and the new born children.

2. **Ozone (O_3)**: Ozone is the major harmful ingredient in the photochemical smog. It is not emitted directly into the air, but is produced in the atmosphere when gases or vapours of organic chemicals called hydrocarbons combine with nitrogen oxide compounds in the presence of sunlight. Ozone is harmful to lungs and decrease their working ability, causing both coughing and chest pains.
3. **Sulphur dioxide (SO_2)**: Sulphur dioxide is produced when sulphur containing fuels are burnt particularly in power plants and diesel engines. This SO_2 convert into sulphuric acid in the atmosphere. SO_2 constricts air passages making it a problem for people with asthma and for young children.
4. **Suspended particulate matter (SPM)**: Particulate matter is the fine particles of dust, which vary from 0.01μm to 0.001 μm. These particulate matters consist of dust, ash and smoke. Burning of coal in the power plant and steel plants is the main source of particulate matter followed by automobiles and stone crushers.
5. **Hydrocarbons (HC)**: These are the organic gases composed of carbon and hydrogen. Insufficient combustion of fuels produces these hydrocarbon gases. These react with nitrogen oxide through photochemical reaction in the presence of sunlight and produce 'smog'.
6. **Carbon dioxide (CO_2)**: Carbon dioxide is the major contributor of air pollution and is released during the combustion of fossil fuels by automobiles and industry. Carbon dioxide is a predominant greenhouse gas, whose concentration in atmosphere has been increasing with rapid industrial growth, and hence its effects in the atmosphere will be experienced for long.

5.5.4 Effects of Air Pollution

Generally, the effects of air pollution can be classified on the basis of duration into two types. Air pollutants show the short term and long term effects on nature and its components.

5.5.4.1 *Short term effects*

Such types of effects can be seen rapidly due to the air pollution. These are as follows:

(a) Visibility reduction: Air pollution is the main cause of visibility reduction. Air pollutants like primary pollutants (coal, smoke, dust particles) and secondary pollutants (photochemical smog) scatter the light and reduce the visibility in atmosphere.

(b) Material damage: Direct damage to structural metals, surface coating fabrics and other materials is a frequent and widespread effect of air pollution. This destruction is related to many types of pollutants, but is chiefly attributed to acid mists, oxidants of various kinds of H_2S and particulate matters of fuel combustion.

(i) H_2S reacts with lead base pigments to blacken it.
(ii) SO_2 affects their composition causing deterioration.
(iii) O_3 is known to cause rapid and extensive damage to many kinds of rubber.

(c) Vegetation damage: In this type of damage, dust and smoke particles fill the leaves pore and reduce the transpiration and evaporation from the plants so that the leaves show chlorosis, glazing, stripping and decreased yield.

(d) **Effects on animals**: Intake air pollutants like heavy metals (arsenic, lead, and acid fumes, which generate from the industries) by the animals results in serve coli, bloody face and a garlic-like odour in the breaths as well as reproductive effects, eyes irritation, fading of hair colour and breathing problems.

(e) **Acid rain**: Acid rain is the form of precipitation with acid, where the water pH becomes low (5.6 to 6.7). It has harmful effects on vegetation, aquatic/terrestrial animals, and infrastructure. Acid rain is mostly caused by emissions of compounds of sulphur (S), nitrogen (N), and carbon (C), which react with the water molecules in the atmosphere to produce acids. However, it can also be caused naturally by the splitting of nitrogen compounds by the energy produced by lightning strikes, or the release of sulfur dioxide into the atmosphere by the phenomena of volcano eruptions. In atmosphere, these sulphur and nitrogen (which come from combustion of fossils and fuels) convert into acid in clouds. Combustion of fuels creates sulphur dioxide and nitric oxides. These are converted into sulphuric acid and nitric acid in the presence of sunlight.

The sequential changes from the emitted gas to acid are as follows:

$$2SO_2 + O_2 \rightarrow 2SO_3$$
$$SO_3 + H_2O \rightarrow H_2SO_4$$

Besides the formation of sulphuric acid, many times in the atmosphere, sulphurous acid is also formed.

$$SO_2 + H_2O \rightarrow H_2SO_3$$

Another cause of acid rain is the oxides of nitrogen, which are released from vehicles and power plants. The sequence of reactions involving oxides of nitrogen in the formation of acid rain is as follows:

$$2NO + O_2 \rightarrow 2NO_2$$
$$4NO_2 + 2H_2O + O_2 \rightarrow 4HNO_3$$
$$2NO_2 + H_2O \rightarrow HNO_2 + HNO_3$$

(1) Effects of Acid rain: Acid rain has adverse effects on forests, freshwater ecosystem, soils, insect and aquatic life-forms. It also causes damage to buildings and has negative impacts on human health.

In soil: Due to acid rain, soil pH decreases and soil biology and chemistry is seriously damaged. Some microbes are unable to tolerate changes due to low pH and are killed. The enzymes of these microbes are denatured (change in shape, so they no longer function) by the acid.

In water: The concentration of aluminium rises in surface water due to acid rain, which causes damage to fish and other aquatic animals. Due to acidification (pH<S) of water streams, eggs of fish get affected (will not hatch in lower pH) and adult fish are killed. However, the extent to which acid rain contributes directly or indirectly via runoff from the catchments to lake and river acidity (i.e., depending on the characteristics of the surrounding watershed) is variable.

Forest and vegetation: Acid rain affects soil and changes its properties, nutrients, minerals converts to lime, sulphate, and nitrate compounds etc. In this way, the soil fertility becomes reduced and it affects the production. Other plants like herbs and crops are also damaged by acid rain due to loss of nutrients.

Human health: Acid rain contains sulphuric and nitric acid, which are harmful to human beings, and scientists have been proved that the same gases as acid rain (sulphur dioxide and nitrogen dioxide), have been shown to cause illness and premature deaths such as from cancer and other diseases.

Effect of acid rain on statues: Acid rain can also causes damage to certain building materials and historical monuments. This results when the sulphuric acid in the rain chemically reacts with the calcium compounds in the stones (limestone, sandstone, marble and granite) to create gypsum, which then flakes off.

$$CaCO_3\ (s) + H_2SO_4\ (aq) \longrightarrow CaSO_4\ (aq) + CO_2\ (g) + H_2O\ ...\ (l)$$

Acid rain also causes an increased rate of oxidation for metals, and in particular copper and bronze. Visibility is also reduced by sulphate and nitrate aerosols and particles in the atmosphere.

5.5.4.2 *Long term effects*

Long term effects take several years for execution, and their effects are seen at global level like global warming and ozone depletion.

(a) **Global warming**: The concept behind it is the rise in temperature simultaneously at global level. The concentration of many gases (CO_2, SO_2, CO, NO_2, CH_4) has increased due to burning of fossil fuels and deforestation, which decrease the surface area available for photosynthesis. These gases in atmosphere absorb the sunlight and raise the temperature of that area. This phenomenon at global level is known as global warming.

(b) **Ozone hole**: In the atmosphere, above 30 km, the stratosphere is present. In the outer shield of stratosphere, the thick layer of ozone is present. This ozone shield absorbs the harmful UV radiation, which comes from the sunlight and prevents it from reaching the surface of the earth. However, air pollutants like NO_2, CFC, SO_2 gases enter into the stratosphere and persist for years until they get converted to other products by reacting with ozone. In this way, Ozone layer becomes thin and this phenomenon is known as O_3 hole. CFCs react with the ozone gas present in the stratosphere and decompose it. Sulphur and nitrogen gases generated from factories, automobiles, etc., also react with O_3 layer and deplete it. These gases can be carried hundreds of kilometers in the atmosphere and break the ozone bond. These gases reaction are given below:

(1) $CFCl_3\ (g) \longrightarrow CFCl_2\ (g) + Cl\ (g);$

$CF_2Cl_2\ (g) \longrightarrow CF_2Cl\ (g) + Cl\ (g);$

$O_3\ (g) \longrightarrow O_2\ (g) + O\ (g);$

(2) $Cl (g) + O_3 (g) \longrightarrow ClO (g) + O_2 (g)$

$O (g) + ClO (g) \longrightarrow Cl (g) + O_2 (g)$

$O (g) + O_3 (g) \longrightarrow 2O_2 (g)$

(3) $O_3 + NO \longrightarrow NO_2 + O_2$

(4) $O_3 + OH \longrightarrow O_2 + \cdot OOH$

$\cdot OOH + O \longrightarrow \cdot OH + O_2$

Due to the air pollution, air pollutants like chlorine reach the stratosphere and react with ozone. Each atom of chlorine can react with more than 10^5 molecules of ozone and convert this ozone into oxygen.

(c) **Photochemical smog**: Photochemical smog is a type of air pollution produced when sunlight acts upon motor vehicle exhaust gases to form harmful substances such as ozone (O_3), aldehydes and peroxyacetyl nitrate (PAN). There are two types of smog: (i) Photochemical smog, (ii) chemical smog.

(i) **Photochemical smog**: The formation of this type of smog is initiated by the UV radiation from sun. The other gradients for the formation of this type of smog are ($NO + NO_2$), unburnt hydrocarbon and SO_2. Oxides of nitrogen and hydrocarbons are produced by the combustion of fuels in automobiles and industries. In the presence of sunlight or UV rays hydrocarbon start a number of chain reactions and that results formation of many secondary pollutants, i.e., smoke, dust, mists, fog, aerosols. This results in the formation of many secondary pollutants (produce combination of smoke, dust, mists, fog, aerosols)

$N_2 (g) + O_2 (g) \longrightarrow 2NO (g)$

$2NO (g) + O_2 (g) \longrightarrow 2NO_2 (g)$

$NO_2 (g) \longrightarrow NO (g) + O$

$O_2 (g) + O \longrightarrow O_3 (g)$

Hydrocarbon + O_3, O_2 [O], $NO_2 \longrightarrow$ Peroxide, peroxyacetyl nitrate (PAN), formaldehyde, ozone aldehyde, acrolein etc.

In the presence of humidity, oxidized hydrocarbons and ozone cause photochemical smog. This is characterized by the formation of aerosols that reduce visibility and generation of brown hazy fumes, which cause irritation in eyes and lungs and also cause extensive damage to vegetation and rubber goods.

(ii) **Chemical Smog**: This type of smog is thought to be caused by sulphur dioxide. It is formed by the combination of smoke, dust, fog and SO_2. Chemical smog is usually formed around some industrial and power plants. It is generally bad in the early morning hours and becomes worse after sunrise due to sunlight induced oxidation of SO_2 to SO_3, followed by reaction with humidity. Sulphuric acid is formed in this type of smog, which is very harmful to people suffering from bronchial diseases such as asthma or emphysema.

Table 5: The sources and effects of gaseous pollutants.

Gaseous pollutants	**Sources**	**Effects**
Carbon monoxide (CO) Nitrous oxide (N_2O)	Home appliances (water heaters, dryers, furnaces, and fireplaces)	Headaches, dizziness, weakness, sleepiness, nausea, vomiting, confusion, and disorientation
Ozone	Automobiles, factories and computer appliances	Shortness of breath, wheezing, coughing, and susceptibility to infections
Formaldehyde	Home and office appliances (carpets, upholstery particle, board and paneling)	Irritation in eyes, nose, throat and headache and respiratory problems
Particulate matter	Particulate pollution arises from many sources and can include dust, wind-blown soil, smoke, soot, and ash. Diesel trucks, coal-driven factories and power plants, wood stoves and agricultural production	Respiratory problems, headache, vomiting, eyes, nasal irritation
Ammonia	Explosive, dye making, fertilizers plants and lacquers	Respiratory problems

5.5.5 Prevention and Control of Air Pollution

There are several techniques by which air pollution can be controlled.

5.5.5.1 *Vegetation*

Air pollution can be reduced by the green belt plan, in which some tree species are selected and planted near the pollution sources like industries, colonies, cement industries, and stone crushers.

5.5.5.2 *Equipment*

Many industries produce various types of gases along with the particulates. To capture these pollutants entering the atmosphere, devices are used depending on the collection capacity and processes used by the particular industry. The following devices are used for dust collection and their removal:

(a) **Cyclone separator units**: These are gravity settling chambers. Cyclone and fabric filters etc. are used to absorb the fine particles in industries. Cyclone is worked to collect relatively large size particulate matter from a gaseous stream through the use of centrifugal forces. Dust laden gas is made to route in a decreasing diameter pathway, forcing solids to the outer edge of the gas stream for deposition into the bottom of the cyclone (Fig. 5.10)

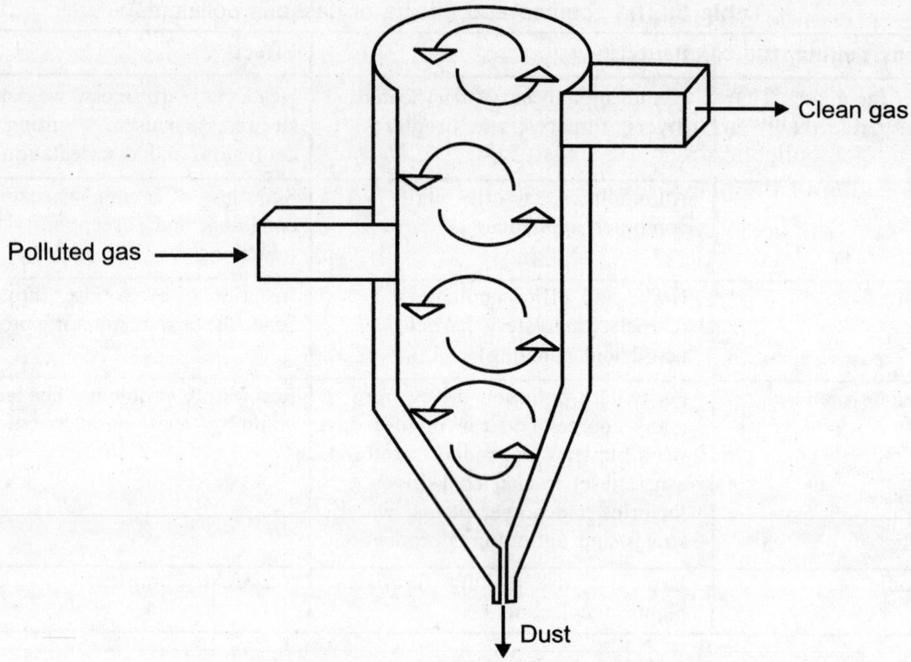

Fig. 5.10: Cyclone separator.

(b) **Wet scrubbing equipment**: Wet scrubbing equipment is the removal of gases and particulate matter from an exhaust stream by dissolving the gaseous contaminants in the liquid stream and by entrapping the solids in the liquid. Gas collection efficiencies in the range of 99% are possible with the wet scrubbers. Fig. 5.11

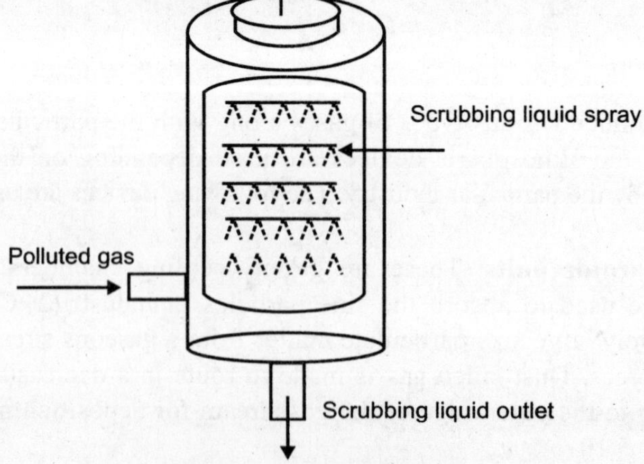

Fig. 5.11: Wet scrubbers.

(c) **Electrostatic precipitator (ESP)**: This device is used for the removal of finest dust particles by the electrostatic attraction that cannot be removed by other equipment such as cyclone separators and fabric filters. This control device utilizes gaseous ions to charge particles, which are then moved through an electric field to be deposited onto charged collection plates. Collected particulate material is then removed by rapping or washing of the plates (Fig. 5.12).

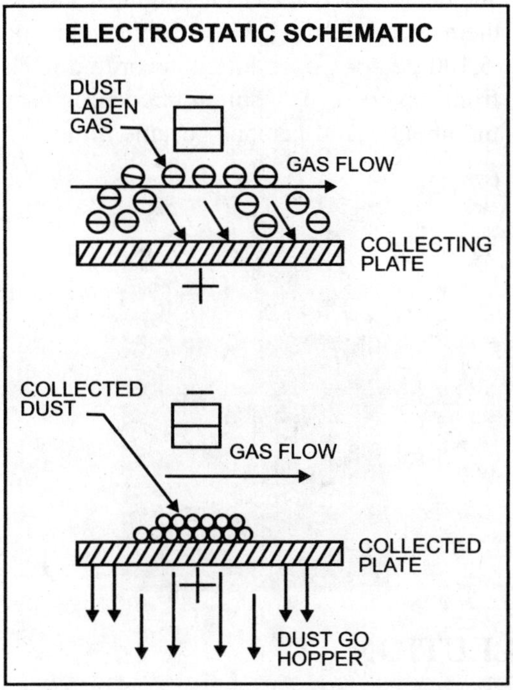

Fig. 5.12: Electrostatic precipitator.

5.5.5.3 *Controlling of air pollution by awareness*

It is must to be aware about pollution and its impacts for its prevention and control. As well as, government should make strict laws and policies for industries. Government agencies like Central Pollution Control Board (CPCB) and State Pollution Control Board (SPCB) should regularly visit the industries and strictly ban such units which are emitting air pollutants or not following CPCB pollution limits.

5.6 CASE STUDY

5.6.1 Bhopal Gas Tragedy

India witnessed one of the most devastating industrial tragedies on 3 December, 1984. This incident took place in Bhopal's Union Carbide plant established by the Union Carbide Corporation of US due to the leakage of methyl isocynate (MIC) The Union Carbide Company knew that there was a problem with the way the plant was functioning, as during the inspection of the plant they noticed that safety measures were not in place and it was not in a good working condition either. And the company's fear of lack of safety measures held fort when on 3rd December 1984, due to an exothermic reaction where water and MIC got mixed led to the gas leaking, and since the safety valve was not functioning, it meant that the gas escaped into the air and started spreading at a rapid rate to the colonies located right next to the plant. And since the incident took place in the night, it claimed even more number of people because most of them were sleeping when the incident took place. The gas spread over 40 km^2 area. About 5,100 persons were killed according to the official records. About 65,000 people suffered from severe eye, respiratory, neuromuscular, gastrointestinal and gynaecological disorder and about 1,000 persons became blind.

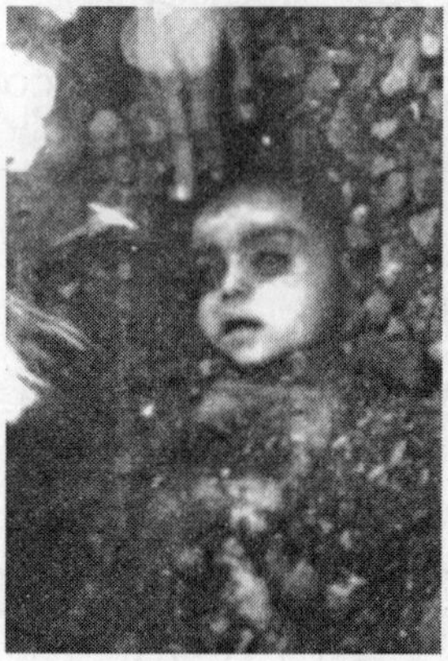

5.7 NOISE POLLUTION

Noise is unwanted sound or sound in a wrong place at a wrong time. It can be defined as unwanted sounds that unreasonably interrupt our daily activities. The definition of noise as unwanted sound implies that it has an adverse effect on human beings and their environment. Noise can also disturb natural wildlife and ecosystem.

Noise sources can be classified into three types like industrial noise, community noise and traffic noise (Fig. 5.13).

5.7.1 Sources of Traffic Noise

5.7.1.1 *Road traffic*

Of all the sources of noise pollution, road traffic is the most prevalent and perhaps the most damaging source of noise pollution. More people are exposed to noise from motor vehicles than any other single source of noise, and thus the impacts of traffic noise are a major factor in human society.

Fig. 5.13: Sources of noise pollution.

5.7.1.2 *Air traffic*

Air traffic is another source of noise pollution. However, airlines travel has become much more common and widely available, resulting in an increased amount of air traffic and thus an increased number of people who are affected by aircraft noise, both passengers and members of communities located near the airports.

5.7.1.3 *Railways*

Railways is the second most important transportation medium after road traffic, not only in India but also in foreign countries. The majority of noise emitted by trains is produced by the engines or by the interaction of the wheels with the track. Other sources of noise in railways include warning signals at crossing, whistles and horns, freight classification yards and railroad construction maintenance equipment.

5.7.2 Sources of Industries Noise

5.7.2.1 *Industry*

All types of industries like heavy and small scale industries also contribute to the noise pollution in the environment. The four main categories of industrial activities that are particularly relevant to the noise are production fabrication, production assembly, power generation and processing. Both people living around the industries and industrial workers are affected by the industrial noise.

5.7.2.2 *Construction*

Construction noise, a major source of noise pollution is emitted by construction equipment. Like industrial equipment, construction equipment tends to produce more noise in the lower end of the frequency spectrum. Construction equipment tend to be used outdoor, and thus affect many other people other than workers at the site.

5.7.3 Sources of Community Noise

Community noise pollution is increasing simultaneously as traffic noise. It includes sirens, agricultural noise, barking dogs, hawker noise, DJ noise, Jagran nights, election campaign and noise generated by humans themselves. Sirens are mostly used by ambulances, police vehicles, and fire engines; sirens can also be considered bothersome, especially by those who live near hospitals, police stations, fire stations.

5.7.4 Effects of Noise Pollution

The affects of noise pollution are of two types:

5.7.4.1 *Physiological*

- Noise pollution affects human health, comfort and efficiency. It causes contraction of blood vessels and makes the skin pale.
- It affects health efficiency and behaviour. It may cause damage to heart, brain, kidney, liver and may produce emotional disturbance.
- Auditory fatigue appears in 90 db associated with whistling and buzzing in ears. Temporary deafness occurs at 4000–6000 Hz, and this effect is known as temporary threshold shift (TTS). Permanent loss of hearing occurs at 100 db due to continuous noise exposure. Under such conditions, the auditory threshold shift is called permanent threshold shift (PTS).
- Besides chronic hearing loss, there may be instantaneous damage or impulsive noise resulting from an explosion or sudden excessive noise of more than 150 db. Impulsive noise also causes psychological and pathological disorders.

5.7.4.2 *Psychological*

- Physiological disorders are also developed due to imbalance in the functioning of body system and due to continuous exposure to noise. These include neurosis, insomnia, hypertension, hepatic diseases, peptic ulcers, and undesirable changes in gastrointestinal activities.
- Ultrasonic sound can affect the digestive, respiratory, cardiovascular systems and semicircular canals of the internal ear.
- Brain is also adversely affected by loud and sudden noise, such as that of jet and aeroplane etc. People are subjected to psychiatric illness.

- The high intensity pulse of sound has its worst effect on the nervous system and it also affects psychomotor performance.

Even the non-living things such as buildings undergo physical damage in the form of cracks, broken windows, door and glasses etc. by sudden and explosive sounds.

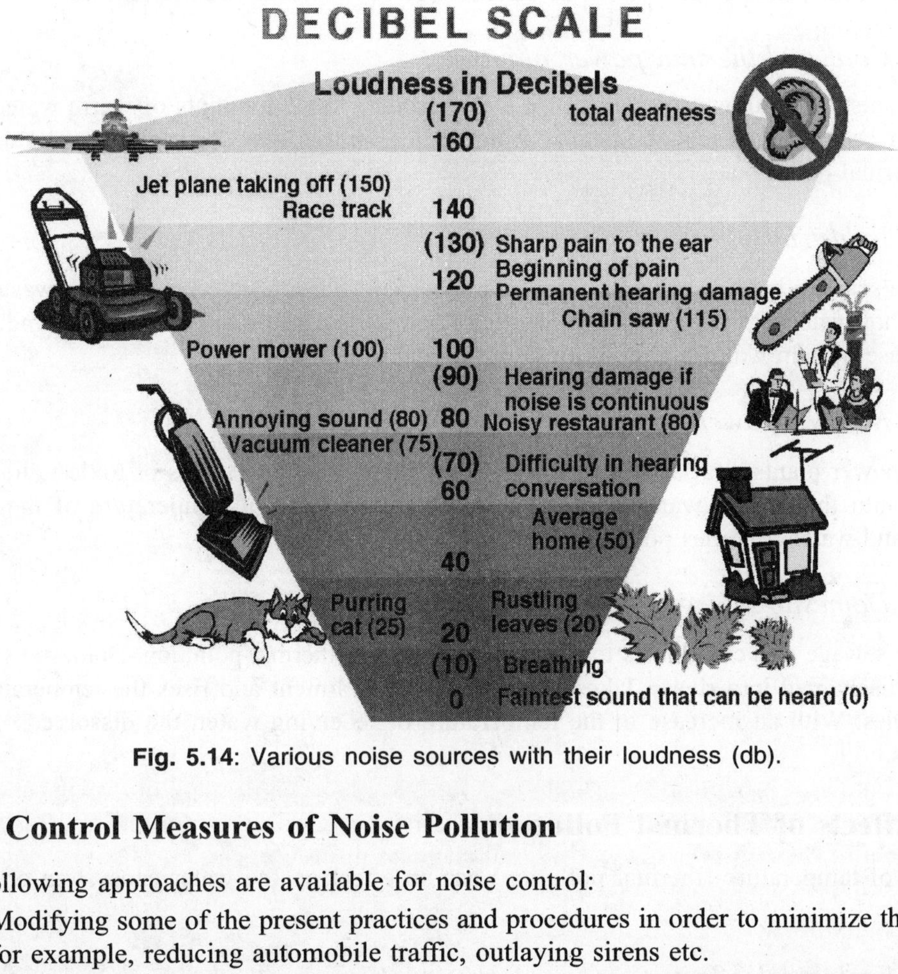

Fig. 5.14: Various noise sources with their loudness (db).

5.7.5 Control Measures of Noise Pollution

The following approaches are available for noise control:

1. Modifying some of the present practices and procedures in order to minimize the noise, for example, reducing automobile traffic, outlaying sirens etc.
2. Shielding the sources of noise generation: Use of sound absorbing motor mounting, better installation design, use of motor enclosures of use vibration damping or absorbing materials in automobiles and dishwashes etc.
3. Shielding the noise receiver: using earplug, ear muffs, helmets, controls booths etc.

5.8 THERMAL POLLUTION

Thermal pollution can be defined as the degradation of water quality by any process that changes the temperature of water. The common cause of thermal pollution is the use of water as a coolant by power plants and industrial manufacturers. However, water (used as a coolant) is returned to the waterbody or natural environment with higher temperature. In this way, the temperature of waterbody rises suddenly, and this high temperature changes the quality of waterbody.

5.8.1 Causes of Thermal Pollution

There are several causes or sources of thermal pollution, which are as follows:

5.8.1.1 *Coal and thermal power plants*

Power plants use coal as a fuel, and as a waste produce large amounts of warm water. This waste hot water dumps into the nearby waterbodies. In this way, such plants pollute water under thermal pollution.

5.8.1.2 *Industrial effluents*

The large scale industries generate their own electricity through the boiler. As a waste they produce hot water. However, some industries like textiles, paper and pulp, sugar industries also release heat in water and contribute in thermal pollution.

5.8.1.3 *Nuclear power plants*

Nuclear power plants emit a large amount of unutilized heat and traces of toxic radioactive material into the nearby water streams, which results in rise of temperature of the water streams and water becomes polluted.

5.8.1.4 *Domestic sewage*

Domestic sewage also contributes to water pollution under thermal pollution. Domestic sewage is often discharged into rivers, lakes, canals without treatment and rises the temperature of waterbodies. With an increase in the temperature of receiving water, the dissolved oxygen decreases.

5.8.2 Effects of Thermal Pollution

The rise of temperature (thermal pollution) has an effect on the water ecosystem or water ecology.

5.8.2.1 *Ecological effects*

Sudden rise in temperature typically decreases the level of dissolved oxygen (DO) in water. The decrease in the level of DO can be harmful for aquatic life (i.e., fish and amphibians).

Thermal pollution may also increase the metabolic rate of aquatic animals, as enzyme activity, resulting in these organisms consuming more food in a shorter time. An increased metabolic rate may create shortage of food sources and has negative effects on aquatic animal population. Changes in the environment may also result in the migration of organisms to another suitable environment. This leads to competition for fewer food resources.

5.8.3 Control of Thermal Pollution

The following methods can be employed to control thermal pollution:

Thermal pollution due to industrial sources is caused mostly by power plants, petroleum refineries, pulp and paper mills, chemical plants and steel mills. Heated water from these sources may be controlled with:

5.8.3.1 *Cooling ponds*

Cooling ponds or reservoirs constitute the simple method of controlling thermal pollution. In this process, first collects the effluents (warmwater) into the blank ponds or pits (cooling pond) for a time and these cooling ponds maximum. This is the simplest method which cools the water to a considerable low temperature.

5.8.3.2 *Cooling towers*

Cooling towers are used frequently to control the thermal pollution. Cooling towers solve the thermal pollution problem by shifting the rejection of heat from waterbodies to the atmosphere, which has a much greater ability to absorb and dissipate the heat input without adverse effects.

In the cooling towers, warm water coming from the condenser is separated downwards over vertical sheets and water flows in thin films. Cool air enters the tower through the water inlet that encircles the base of the tower and rises upward, causing evaporative cooling. A natural air flow is maintained because of the density difference. Heat is dissipated to the atmosphere about 100 m above the base of the tower and recycled back to the power plant condensers Fig. 5.15.

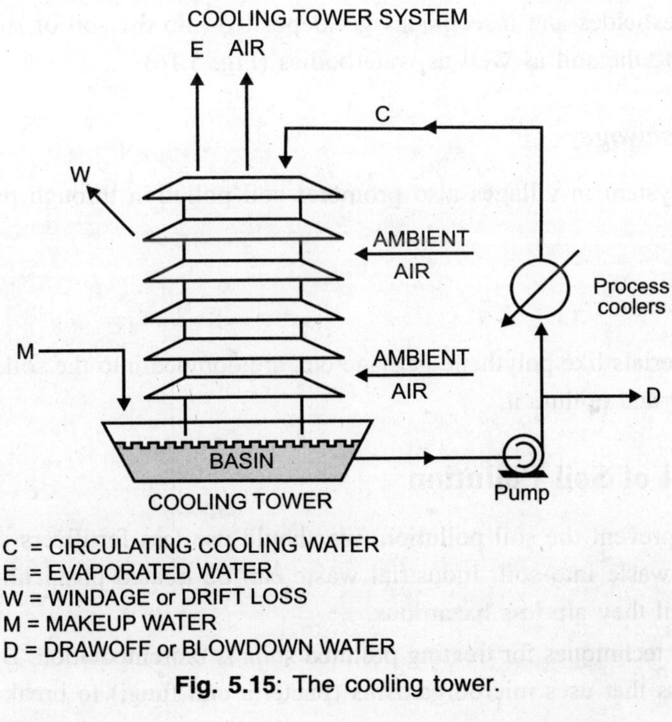

Fig. 5.15: The cooling tower.

5.9 SOIL POLLUTION

Any part of the earth's surface that supports vegetation or root is called soil. Soil pollution is the introduction of substances, biological organisms or energy into the soil, resulting in a change of soil quality, which is likely to affect the normal use of soil and living environment. In simple words, soil pollution can be defined as the reduction in the soil fertility and quality.

5.9.1 Sources of Soil Pollution

There are many direct and indirect sources which pollute the soil.

5.9.1.1. *Artificial fertilizers*

Soil nutrients are important for plant growth and development. Plants obtain carbon, hydrogen and oxygen from air and water, but other necessary nutrients like nitrogen, phosphorus, potassium, calcium, magnesium, and sulphur must be obtained from the soil. Farmers generally use fertilizers to improve the soil deficiency for crop production. Sometimes these artificial fertilizers, which are formed by the hazardous chemicals or stay for long into the soil and disturb the soil quality parameters, enter the food chain.

5.9.1.2 *Pesticides, herbicides and insecticide*

Pesticides and insecticides are used by farmers in their fields to kill unwanted insects and weeds. These pesticides and insecticides get deposited into the soil or run-off through rain water and pollute the soil as well as waterbodies (Fig. 5.16)

5.9.1.3 *Open sewage*

Open sanitary system in villages also promotes soil pollution through pathogenic bacteria and fungi.

5.9.1.4 *Solid wastes*

Solid waste materials like polythene garbage etc. are dumped into the soil, which reduce the soil productivity and pollute it.

5.9.2 Control of Soil Pollution

To control and prevent the soil pollution, we should use less fertilizers and pesticides and dump the solid waste into soil. Industrial waste can be treated physically, chemically and biologically, until they are less hazardous.

One of the techniques for treating polluted soils is bioremediation. Bioremediation is a treatment process that uses microorganisms (bacteria and fungi) to break down or degrade hazardous substances into less toxic or non-toxic substances. The favorable conditions for bioremediation are:

- Temperature for organism (25°C – 30°C)
- Availability of moisture (70%–80%)
- Composition of nitrogen, potassium and phosphorus (3:2:1)
- C: N ratio (2:3)

Availability of oxygen in sufficient quantity in the soil (not less than 5 ppm)

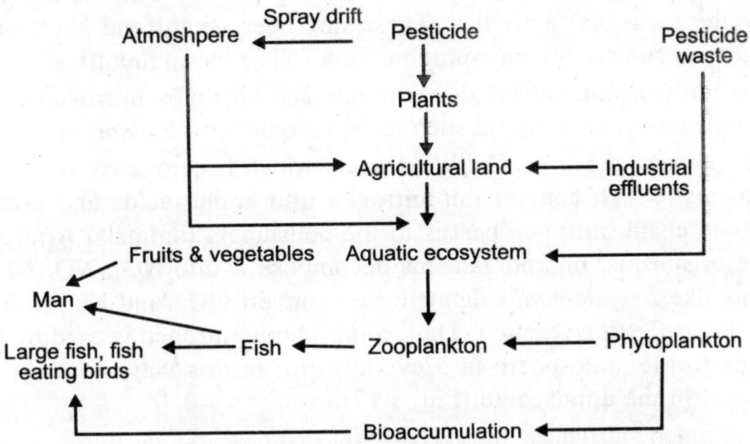

Fig. 5.16: Possible pathways and sources of pesticides.

5.10 BIOGEOCHEMICAL CYCLES

The chemical elements tend to circulate in the biosphere in characteristic path from environment to organisms and back to the environment or nature. These more or less circular paths of the chemical elements are known as biogeochemical cycles. These cycles of chemical elements involve biological organisms and their geological environment. Hence, they are collectively referred to as biogeochemical cycles. In this term, "bio" refers to living organisms and "geo" refers to the rocks, soil, air and water of the earth.

There are two basic types of biogeochemical cycles:

1. **Gaseous types**: In these types of biogeochemical cycles, atmosphere constitutes the major reservoir of the element, which exists there in gaseous phase. The carbon, oxygen, and nitrogen cycles are good examples of gaseous type of biogeochemical cycle.
2. **Sedimentary types**: In these types of cycle, major reservoir is the lithosphere from which the elements are released by weathering. The phosphorous and sulphur cycles are best examples of sedimentary types of biogeochemical cycle.

5.10.1 Gaseous Cycle

5.10.1.1 *Nitrogen cycle*

The atmosphere is composed of several gases, out of which nitrogen (N_2) is 79% in the atmosphere. It is required by organisms in the synthesis of proteins, nucleic acid and other nitrogenous compounds. Atmospheric nitrogen serves as the ultimate source, but it cannot be used by living organisms directly.

In the nitrogen cycle, the atmospheric nitrogen is converted into its compounds like nitrates (through thundering lightning and rainfall) In soil, this nitrogen is present in the form of ammonia (NH_4), nitrate and nitrite, which are generated from the organic nitrogenous substances by the microbial activities. These microbes (fungi and bacteria) convert the organic nitrogenous substances into ammonia ion (NH_4) by ammonification reaction.

Ammonia ion is again converted into nitrate and nitrite by nitrification reaction. This reaction is completed by the bacteria such as Nitrosomonas, *Nitrosococcus*.

After this process, the atmospheric nitrogen, which is converted to NO_2 and NO_3 ions used by the plants, which convert this nitrogen into amino acids and proteins as food. Through the food chain, nitrogen passes to the consumers (animals) After the death and decay of these organisms, microorganisms decompose it into NO_2, NO, NH_4 ions. Some microorganisms like *Pseudomonas* denitrificans convert NO_2, and NO_3 into N_2, N_2O and NO_2 gases by denitrification reaction. Thus, atmospheric nitrogen is used by the organisms and return back to the atmosphere in a cyclic form. In this way, nitrogen concentration becomes constant in the atmosphere (Fig. 5.17).

Organic nitrogen substances $\longrightarrow NH_4, NO_3, NO_2$

$2NO_3 \longrightarrow NO_2 + 2O_2$

$2NO_3 \longrightarrow 2N_2O + 5O_2$

$2NO_3 \longrightarrow 2NO + 2O_2$

$2NO_3 \longrightarrow N_2 + 3O_2$

5.10.1.2 *Oxygen cycle*

Atmosphere contains about 21% O_2. The atmospheric oxygen enters the living organisms as a gas required in respiration. During this process, CO_2 and H_2O are formed.

$$C_6H_{12}O_6 + 6O_2 \rightarrow 6CO_2 + 6H_2O + \text{Energy}$$

The metabolic water thus formed is added to all other water present in living organisms, and as such it may undergo three possible fates. Some of it may be excreted and so added to the water content of the environment. Another part may be used as a building material for the formation of more living matter. A third possible fate of the water within the organism is its consumption as a fundamental raw material along with the carbon dioxide in photosynthesis.

$$6CO_2 + 6H_2O + \text{light} \rightarrow C_6H_{12}O_6 + 6O_2$$

Such free oxygen may now be again used in respiration, or it may be returned to the atmosphere or nature as molecular atmospheric oxygen, thus completing the cycle. Thus, oxygen enters the organisms only through respiration and leaves through photosynthesis (Fig. 5.18).

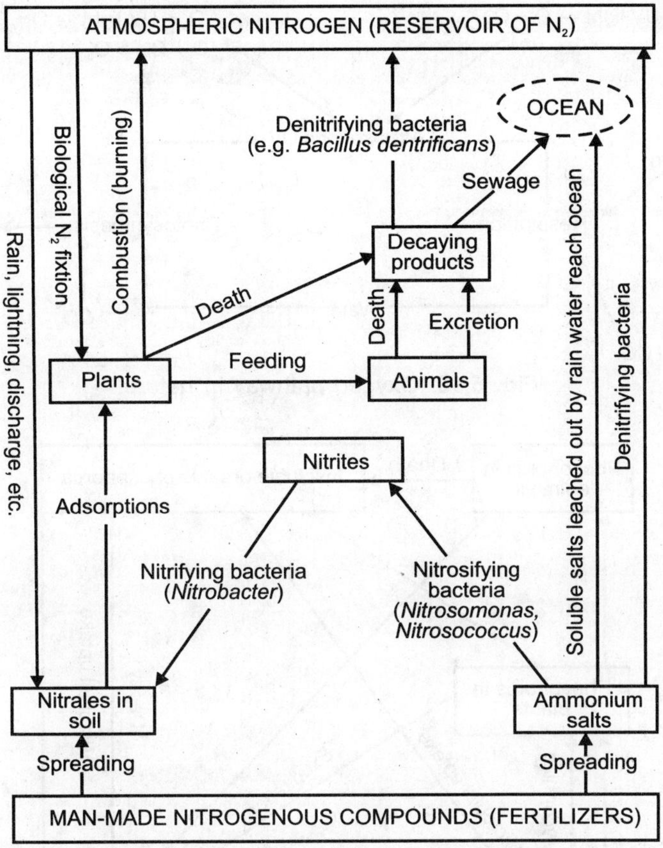

Fig. 5.17: Nitrogen pathway in nature.

5.10.1.3 *Phosphorus cycle*

The basic source and the great reservoir of phosphorus are the rocks and other deposits which have been formed in the past geological ages. But much phosphate escapes into the sea, where part of it is deposited in the shallow sedimentations, and a part of it is lost to the deep sediments.

The fish and marine birds are also important in phosphorus cycle. The latter have apparently played an important role in returning phosphorus to the cycle. Although we harvest a lot of fish, about 60,000 tonnes of elementary phosphorus per year is returned in this manner. Plants absorb water dissolved inorganic phosphate from the soil either as dihydrogen phosphates (H_2PO_4) or (HPO_4) and convert it into organic phosphate. The latter is transferred to animals and decomposers. Decomposers return phosphorus to the soil as phosphate ions. Phosphate absorbed from the soil is returned to it in dead plants and animals' organic residues, which are converted to humus by the action of soil microbes. Thus, phosphate returns back to the lithosphere, and completes the cycle (Fig. 5.19)

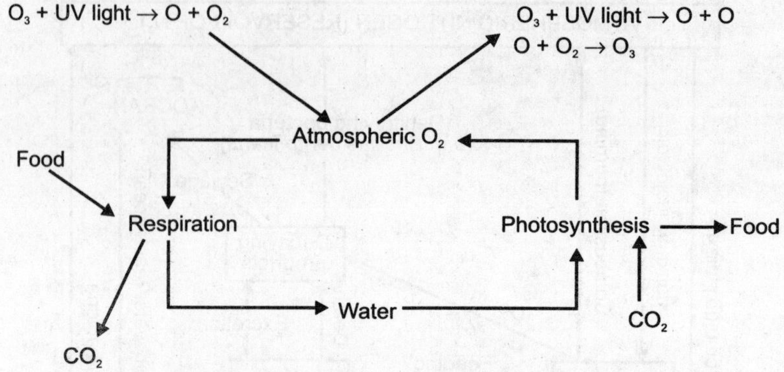

Fig. 5.18: Oxygen pathway in nature.

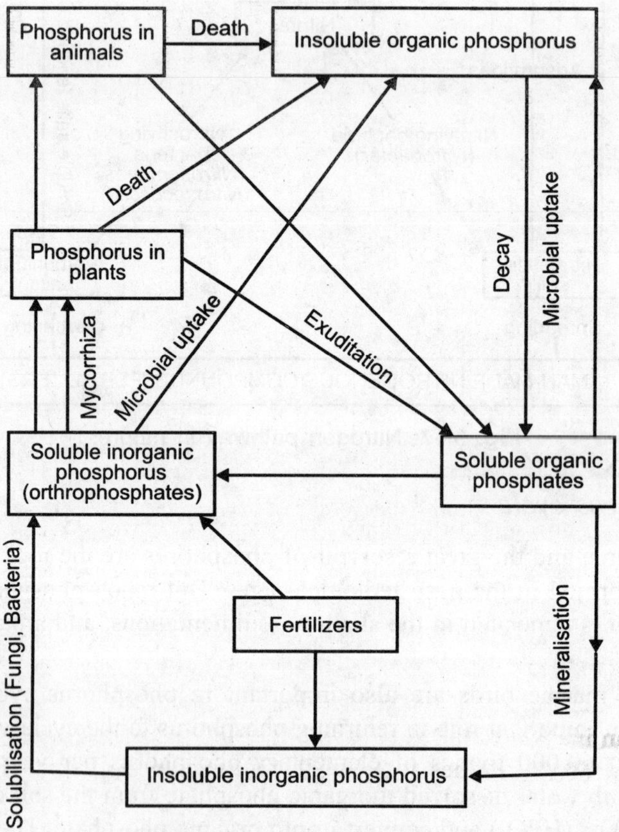

Fig. 5.19: Phosphorus pathway in nature.

5.11 HAZARDOUS WASTE

Different types of materials are produced as a result of different types of manufacturing processes. These wastes are very toxic and for the environment and people, and are called hazardous waste.

5.11.1 Types of Hazardous Wastes

There are many types of hazardous wastes depending on their dangerous aspects. However, these are very harmful for the environment. Some of the hazardous wastes are as follows:

- Nuclear waste
- Biomedical waste
- Industrial waste
- Universal waste

5.11.1.1 *Nuclear waste*

It is produced by nuclear plants and radioactive materials. Nuclear wastes cause different types of diseases and the destructive problems.

5.11.1.2 *Biomedical waste*

The other type of hazardous waste is biomedical waste. Biomedical wastes are generated from the insecure disposal of medical products. Different types of medical products such as hospital waste. This waste is hazardous in nature and causes incurable diseases.

5.11.1.3 *Industrial waste*

The industrial waste is generated as a result of different types of manufacturing processes such as the manufacture of new products for the benefit of mankind. However, this problem is generated from the chemicals that are used in the production of different types of goods, like varnishes, wood preservation chemicals and medicine, etc.

5.11.1.4 *Universal waste*

Another type of hazardous waste is universal waste. These wastes are generated by the products that are used both commercially and in personal life. Some important products that producethe universal waste are light bulbs, batteries, electronics goods etc.

5.11.2 Control of Hazardous Waste

Hazardous waste is a severe problem. Some of the methods to control the hazardous waste are given below:

5.11.2.1 *Landfill*

Landfill is a common approach to dispose the waste. In this method, waste is buried in the wasteland. Landfills were often established in unused quarries, mining voids or borrow pits (Fig. 5.20) A properly designed and well-managed landfill, i.e. aerobic & anaerobic bacteria, algae, fungi and worms, which is made of clay and plastic lining material. Deposited hazardous waste is normally compacted to increase its density and stability and covered to prevent attracting pests, rats, and cockroaches. It is an inexpensive method of disposing of waste materials. However, poorly designed or managed landfill sites create a number of adverse environmental effects such as attracting the pests, rats, cockroaches, and polluting the ground water due to the percolation of waste. Another common byproduct of landfills is gasses like CH_4 and CO_2, which are produced during decomposition by the anaerobic microbes.

5.11.2.2 *Incineration*

Incineration is another disposal method in which hazardous (biomedical and chemical) and solid organic wastes are combusted to convert them into residue and gaseous products. The incineration process reduces the volumes of solid waste up to 20 to 30 percent of the original volume. The incineration and other high temperature waste treatment processes are also known as "thermal treatment" (Fig. 5.21) In this process, waste materials are converted into non-pollutant products like heat, gas, steam and ash. Nowadays, incineration is carried by both small and large-scale industries. It is also used to dispose of all types of solid, liquid and gaseous waste.

5.11.2.3 *Recycling*

Recycling is also an important method to control the hazardous waste. In this method, first the waste in collected and reused (i.e., beverage containers) The most common consumer products recycled include aluminium such as beverage cans, copper such as wire, steel food and aerosol cans, old steel furnishings or equipment, polyethylcnc and PET bottles, glass bottles and jars, paperboard cartons, newspapers, magazines and light paper, and corrugated fiberboard boxes. These items are usually composed of a single type of material, making them relatively easy to recycle into new products. However, complex products like computer and electronic equipment are more difficult to recycle, although nowadays electronic goods manufacturing companies use recycling material in their products.

5.11.2.4 *Biological reprocessing*

Organic waste materials in nature, like plants, waste food, and paper products are used, as a food for microorganisms (aerobic and anaerobic bacteria) and microorganisms decompose and digest this organic matter. This matter gets converted into compost, which can be used for agricultural and landscaping purposes. In addition, methane gas is also produced in this process, can be used for generating electricity and heat. The intention of biological processing in waste management is to control and accelerate the natural process of decomposition of organic matter.

Fig. 5.20: The landfill sites of nuclear hazards waste.

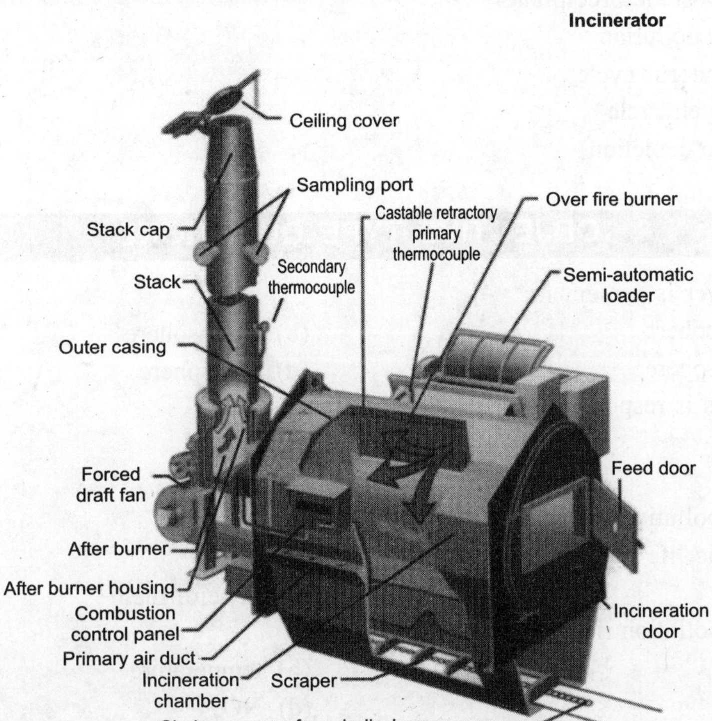

Fig. 5.21: The incinerator which is used to incinerate the hazardous waste.

EXERCISE

QUESTIONS

1. Define environmental pollution, its impact and mitigation.
2. Name and explain various types of pollution.
3. Differentiate between pollutant and pollution.
4. What are biogeochemical cycles?
5. What are the sources of air pollution and their control?
6. Explain thermal pollution and its control.
7. What is soil pollution? How can it be controlled?
8. Explain acid rain and photochemical smog and their prevention.
9. Describe the effluent treatment plant of any industry.
10. Write short notes on:
    - BOD
    - COD
    - Eutrophication
    - Primary and secondary treatment unit
    - Electrostatic precipitator
    - Noise pollution
    - Phosphorus cycle
    - Nitrogen cycle
    - Ozone depletion

OBJECTIVE TYPE QUESTIONS

1. Ozone layer is present in:
   (a) troposphere (b) stratosphere
   (c) mesosphere (d) ionosphere
2. Which gas is responsible for ozone depletion?
   (a) CFC (b) SO_2
   (c) H_2O (d) NO_2
3. Thermal pollution has impact on:
   (a) aquatic life (b) soil
   (c) both (d) none of them
4. Thermal pollution rises:
   (a) O_2 (b) temperature
   (c) H_2O (d) NO_2

5. Aeroplane flies on which layer of atmosphere?
   (a) Troposphere (b) Stratosphere
   (c) Mesosphere (d) Ionosphere
6. Air pollution has impact on:
   (a) plants (b) animals
   (c) both (d) none of them
7. Noise pollution is measured in
   (a) decibel (b) dobson
   (c) decimal (d) ppm
8. Bhopal gas tragedy occurred on:
   (a) 3 December, 1984 (b) 5 December, 1984
   (c) 4 December, 1994 (d) None
9. The gas responsible for Bhopal gas tragedy is:
   (a) methane (b) methyl isocyanide
   (c) CO (d) methyl propane
10. BOD means:
    (a) biological oxygen demand (b) biological oxygen diamond
    (c) both (d) none
11. Components of acid rain are:
    (a) H_2SO_4 (b) HCL
    (c) HNO_3 (d) all
12. Photochemical smog mainly contains:
    (a) CFC (b) SO_2
    (c) H_2O (d) NO_2
13. Greenhouse gases are:
    (a) CH_4 CO_2 (b) NO_2 CO_2
    (c) CO CO_2 (d) SO_2 NO_2
14. Mercury is responsible for:
    (a) minamata (b) saphogation
    (c) dehydration (d) all
15. Fluorosis disease is caused due to:
    (a) fluoride (b) fericoxide
    (c) both (d) iodine
16. Eutrophication is caused by:
    (a) overnutrition (b) SO_2
    (c) H_2O (d) NO_2
17. COD means:
    (a) chemical oxygen demand (b) chlorine oxygen demand
    (c) both (d) none

18. Causes of marine pollution are:
(a) refineries (b) industries
(c) domestic waste (d) none
19. BMW refers to:
(a) British Motor Work (b) Bharat Motor Work
(c) Biomedical waste (d) None
20. The point source of water pollution is:
(a) STP (b) fertilizer run-off
(c) pesticides (d) none
21. Non-point water pollution source is:
(a) nutrition run-off (b) industries
(c) domestic waste (d) oil spills
22. Commonly pollutants are of:
(a) 2 types (b) 3 types
(c) 4 types (d) more than 4 types
23. Air pollution is of:
(a) 2 types (b) 3 types
(c) 4 types (d) none of them
24. PAN means:
(a) para acetyl nitrate (b) para amide nitrate
(c) both (d) none
25. Chemical smog is caused by:
(a) SO_2 (b) NO_2
(c) methane (d) ethane
26. The main source of soil pollution is:
(a) chemical fertilizer (b) pesticides
(c) herbicides (d) all
27. What are the pollutants of underground water?
(a) pesticides (b) biomedical waste
(c) chemical fertilizer (d) all
28. Gaseous type of geochemical cycle is:
(a) NO_2 (b) SO_2
(c) CO_2 (d) all
29. Sedimentary type of geochemical cycle is:
(a) phosphate (b) sulphur
(c) methane (d) all
30. Ecological effects of thermal pollution are:
(a) decreased DO level (b) decrease CO_2 level
(c) decreased temperature (d) all

31. Noise pollution can be controlled by:
    (a) plantation (b) ear plugs
    (c) sound proof instruments (d) none
32. In nitrogen cycle, organic nitrogen converts into ammonia (NH_4) by a reaction known as:
    (a) ammonification (b) acidification
    (c) oxidation (d) reduction
33. Nitrification reaction in nitrogen cycle is completed by:
    (a) Nitrosococcus bacteria (b) Bacillus
    (c) Nitrobacillus (d) Nitro comma bacteria
34. Ammonia ion converts into nitrate and nitrite by a chemical reaction known as:
    (a) denitrification (b) nitrification
    (c) ammonification (d) all
35. Reaction in which nitrite is converted into atmospheric nitrogen is known as:
    (a) nitrification (b) denitrification
    (c) reduction (d) oxidation

ANSWERS

1. (b) 2. (a) 3. (a) 4. (b) 5. (a) 6. (c) 7. (b) 8. (a) 9. (b) 10. (a)
11. (d) 12. (b) 13. (a) 14. (a) 15. (a) 16. (a) 17. (a) 18. (a) 19. (c) 20. (a)
21. (a) 22. (d) 23. (b) 24. (a) 25. (a) 26. (d) 27. (d) 28. (d) 29. (a) 30. (a)
31. (a) 32. (a) 33. (a) 34. (b) 35. (b).

6

ENERGY RESOURCES

6.1 INTRODUCTION

Energy is a primary input for almost all economic activities and is, therefore, vital for improvement in the quality of life. The standard of living of the people of any country is considered to be proportional to the energy consumption by the people of that country. In one sense, the disparity one feels from country to country arises from the extent of accessible energy for the citizens of each country. The world population will increase from 6 billion to 11 billion in this century, and the life expectancy has increased 2 times in the last two centuries and the energy requirement has increased by 35 times in the same period. The main drivers of the alternate energy search are the population growth, economy, technology, and agriculture. Unfortunately, the world energy demand is increasing and energy resources are becoming scarce and costlier. The demand of energy can be accomplished by the use of energy sources.

A source of energy is one that can provide adequate amount of energy in a usable form over a long period of time. These sources can be of two types:

- **Renewable Energy Resources**: Renewable energy sources can be regenerated continuously in nature, or in other words, the energy sources that can regenerate within a short period are known as renewable energy sources, e.g., wood, solar energy, wind energy, tidal energy, hydropower, biofuels and geothermal energy. These are also known as non-conventional sources of energy and can be used again and again.
- **Non-Renewable Energy Sources**: Such energy sources accumulated in nature over a long time and cannot be quickly replenished when exhausted. In other words, such energy sources take maximum time for renovation or regeneration e.g., coal, petroleum, natural gases and nuclear fuel.

6.2 RENEWABLE ENERGY RESOURCES

6.2.1 Solar Energy

Sun is the ultimate source of energy, directly and indirectly for all other forms of energy. Solar energy is derived from the sun in the form of solar radiation. Solar powered electrical generation relies on photovoltaic and heat engines. A partial list of other solar applications includes space heating and cooling through solar architecture, day lighting, solar hot water, solar cooking, and high temperature process heat for industrial purposes. Solar technologies are broadly characterized as either passive solar or active solar depending on the way they capture, convert and distribute solar energy. Active solar techniques include the use of photovoltaic panels and solar thermal collectors to harness the energy. Passive solar techniques include orienting a building to the sun, selecting materials with favourable thermal mass or light dispersing properties, and designing spaces that naturally circulate air.

6.2.2 Hydropower Plants

Hydroelectric energy is generated from the kinetic energy of water. Water falls continuously from the height (dams) into the turbines (situated at the bottom), and this turbine converts water kinetic energy into electric energy. Approximately one-fourth of the world's electricity is produced by hydropower.

6.2.2.1 *Advantages of hydropower*

1. It is cheaper than the electricity produced by the thermal plants.
2. Hydropower plants do not pollute the environment.
3. It usually exploits the naturally available energy potential of water.
4. Keeping disaster management in view, mini hydel plants are established in hilly areas, which provide electricity to local areas.

6.2.3 Wind Energy

Wind is another renewable energy. Wind power can be used in certain areas on certain days. Wind power is used to run the turbine system and produce electricity. In India, electricity generation from the wind power sector is only 3% of the total installed wind mills.

6.2.3.1 *Advantages of wind power*

1. The cost of production is not high; it can be reduced through research and better management.
2. The operation and maintenance costs are not high.
3. Land below the turbines can be utilized for growing crops and stock grazing.
4. It produced more energy due to high efficiency.

6.2.4 Tidal Energy

The kinetic energy of tidal waves is used to run the turbine and generate electricity (Fig. 6.1a & b). The tidal energy can be harnessed by constructing a tidal barrage. During the high tide, the sea water flows into the reservoir of the barrage. This water flow turns the turbine to produce electricity. Under the low tide, water from the reservoir flows back into the sea and turns the turbine to generate electricity.

Fig. 6.1: **(a)** The wind mills; **(b)** The movement of turbine with tidal waves.

6.2.5 Biofuels

Biofuels are alternate source of energy, and a wide range of fuels are obtained from biomass (bioethanol, biodiesel). The term covers solid biomass, liquid fuels and various biogases. Bioethanol is an alcohol-made fuel which is obtained by the fermentation of sugar components of plant materials (i.e., sugarcane and starch crops). Today, by using the advanced technology, biofuels are being developed by cellulose biomass (i.e., trees and grasses) and feedstocks are used for ethanol production. Ethanol is an alternate fuel for vehicles in its pure form. Many developed countries (USA) are using bioethanol widely as a fuel. Another biofuel known as biodiesel is made from vegetable oils, animal fats or recycled greases. Biodiesel is also used as a fuel for vehicles in its pure form, however, it is usually used as a diesel additive to reduce the levels of particulates, carbon monoxide, and hydrocarbons from diesel-powered vehicles. It is the most common biofuel in advance countries (Europe).

6.2.5.1 *Differences between biofuels and fossil fuels*

- Fossil fuels take millions of years to produce whereas the biofuels can be made extremely fast as compared to fossil fuels.
- Huge amounts of air pollution are generated by fossil fuels as compared to biofuels.
- Unlike fossil fuels, biofuels are renewable sources of energy.

6.2.5.2 *Different types of biofuels*

There are four types of biofuels:

(a) **First generation biofuels**: The sources from where the first generation biofuels are derived are vegetable fats, starch, and sugar. They are also derived from food-crops.

The first generation fuels are also derived from animal fats. Some examples of this type of biofuels are biogas, biodiesel, and vegetable oil.

(b) **Second generation biofuels**: The source of second generation biofuels is derived from waste biomass. Thus, this thing makes it a more balanced option as compared to the first-generation biofuels. The examples of second generation biofuels are different kinds of alcohols and diesel generated from wood.

(c) **Third generation biofuels**: The sources of third generation biofuels are derived from algae and plants (Fig. 6.2 (a & b)). On a large scales, algae are farmed for creating these biofuels. As algae fuels can easily decompose into the soil without harming it, therefore they are extremely environment-friendly.

(d) **Fourth generation biofuels**: To derive the biofuels of the fourth generation, a method is used in which microorganisms are raised to work with carbon dioxide for the purpose of generating fuel.

6.2.5.3 *Advantages of biofuels*

- Biofuels are alternate source of fossil fuels.
- Biofuels are environment-friendly. They are helpful in reducing carbon emissions into the atmosphere.

It has been proven that biofuels, especially, biodiesel is very cost-effective.

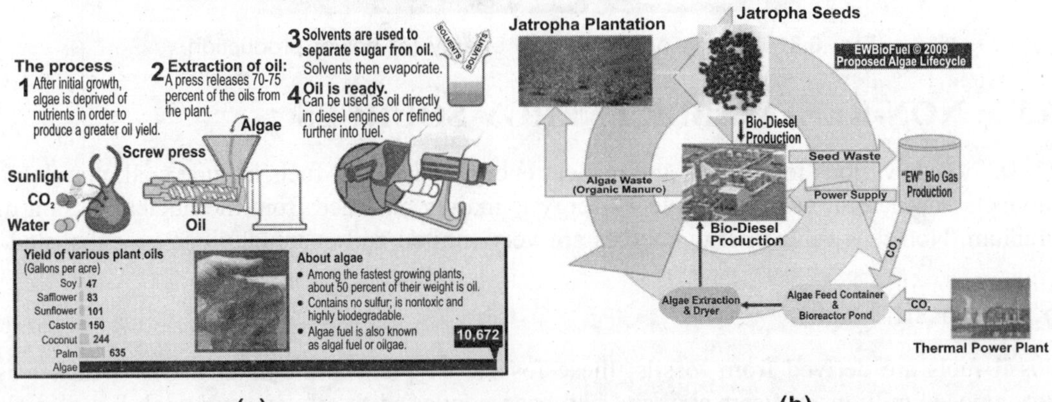

Fig. 6.2 (a): The formation of biofuel through algae; **(b)** The formation of biofuel through Jatropha.

6.2.6 Geothermal Power

Geothermal power means earth heat (Greek *geo*, meaning earth, and *thermos*, meaning heat). A large amount of heat is stored in the earth to extract this heat and utilization called geothermal energy. The geothermal energy originates from the origin of the planet, radioactive decay of minerals, and solar energy absorbed at the surface. Geothermal energy has been used for bathing since Paleolithic times; it is now used for electricity production (Fig. 6.3). Geothermal power is cost effective, reliable, sustainable, and eco-friendly, however this energy can be obtained from near tectonic plate boundaries. Advanced technologies have dramatically expanded the range and size of viable resources, especially for applications such as home heating, opening a potential for widespread exploitation. Although, geothermal wells release greenhouse gases from the earth, but these emissions are very low per energy unit production than those of fossil fuels. Due to this, it can be said that, geothermal power has the potential to help mitigate global warming if widely deployed in place of fossil fuels.

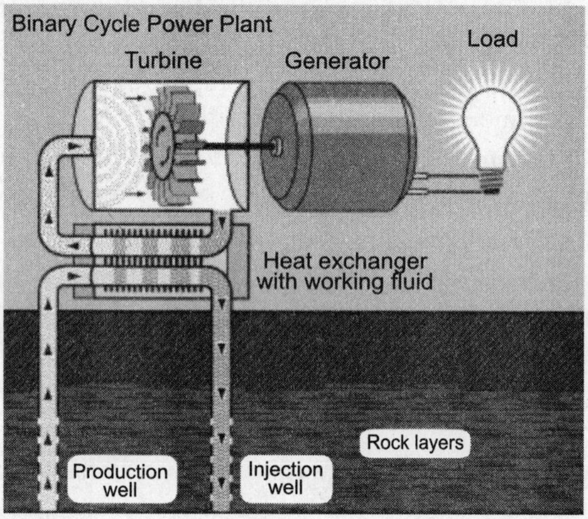

Fig. 6.3: The use of geothermal energy for power production.

6.3 NON-RENEWABLE ENERGY SOURCES

These include various fossil fuels and nuclear energy. The fossil fuels include coal, petroleum products and natural gas. The nuclear energy is mainly obtained from the nuclear fission of uranium. Non-renewable energy sources are very limited and eventually can be exhausted.

6.3.1 Fossil Fuels

Fossil fuels are derived from fossils; these fossils are plant and animal matter which was decomposed millions of years ago, and had been converted to oil, gas, and coal. Fossil fuels take a very long time to form, therefore these are called non-renewable sources. After millions of years, decomposed plants and animals get converted into a hard, black rock-like substance called coal, a thick liquid called oil or petroleum, and natural gas. These are the three major forms of fossil fuels. These are usually found underground.

Coal is generally collected through either mining or dug out, while oil and natural gas are pumped out from the ground. Coal is widely distributed on the earth and is easier to locate than oil and petroleum gas. Fossil fuels take millions of years to form, but burn and disappear in a few seconds. However, due to lack of awareness, people have irretrievably damaged the planet by extracting and burning these fuels. These fossil fuels should not be wasted as these are non-renewable. Every year, millions of tonnes of coal is consumed. This has led to global warming (greenhouse effect) and depletion of resources.

6.3.2 Coal

Coal is the most abundant fossil fuel on the earth. The main element of coal is carbon, which is mainly used as a combustion fuel. The large-scale use of coal began with the Industrial Revolution in the 19th century. As the number of industries increased, demand for more sources of energy grew. Coal is the product of plants, mainly trees, which had died millions of years ago. Due to water logging in low-lying swampy areas or in slowly sinking lagoons, dead trees and plants did not decompose as they normally would. The dead plant matter was covered with water and protected from the oxidizing effect of air. The action of certain bacteria released the oxygen and hydrogen, making the residue richer and richer in carbon. Thick layers of this carbon-rich substance, called peat, was built up over thousands of years. As more material accumulated above the peat, the water was squeezed out leaving just carbon-rich plant remains. Pressure and temperature further compressed the material. This aided the process of producing coal as more gases were forced out and the proportion of carbon continued to increase. The carbon slowly metamorphosed into coal over millions of years. There are three main types of coal: lignite, bituminous, and anthracite. Lignite and bituminous have lesser percentage of carbon, and they burn faster. Anthracite has about 98% carbon, therefore it burns slowly and releases less smoke. All types of coal contain sulphur. Sulphur is the worst of the pollutants and causes damage to human health and vegetation (Fig. 6.4a & b). Coal is the principal heat source for electricity generation in most of the countries, and is used directly in such heavy industries as iron and steel making. Coal comes from underground mines. Today almost 80% of the coal comes from surface strip mines (opencast mines), which is much safer.

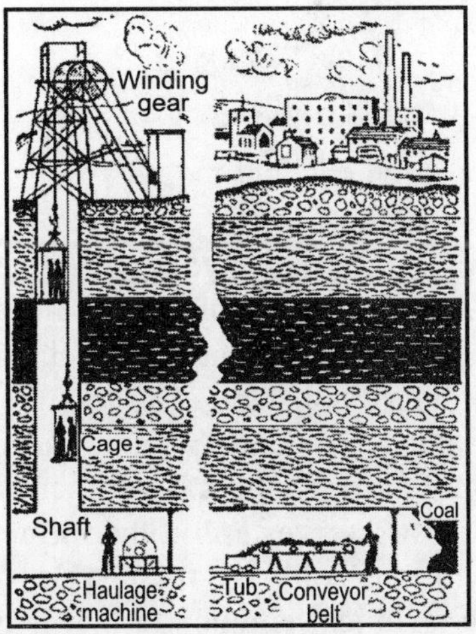

(a)

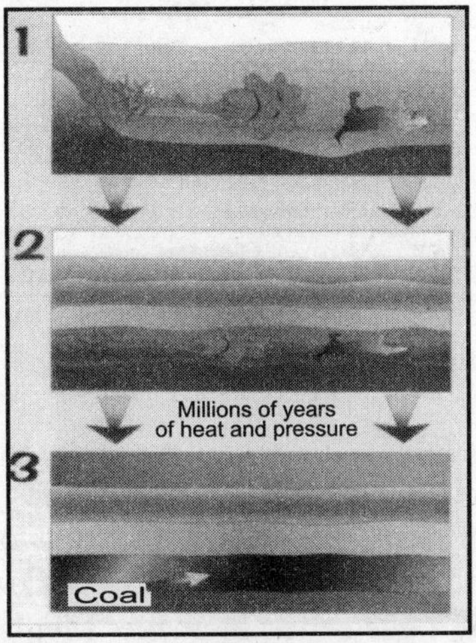

(b)

Fig. 6.4 (a): The coal mine; **(b)** The formation of coal.

6.3.3 Oil and Gas

Almost all oil and natural gas are found deep underground in tiny holes in rocks. Millions of years ago, sea covered much of what is now dry land. In prehistoric times, tiny plants and animals lived in the sea. When these creatures died, they sank to the bottom of the sea, and got buried in layers of mud and sand. As the times passed, this organic material sank deeper and deeper (Fig. 6.5). The earth's crust changed its shape, and put intense pressure and heat on plants and tiny animals. Heat from the earth's interior and the weight of the overlying rocks gradually changed the energy-containing substances in the accumulated plants into hydrocarbon liquids and gases. As millions of years passed, these deposits turned into chemicals, which are called 'hydrocarbons'.

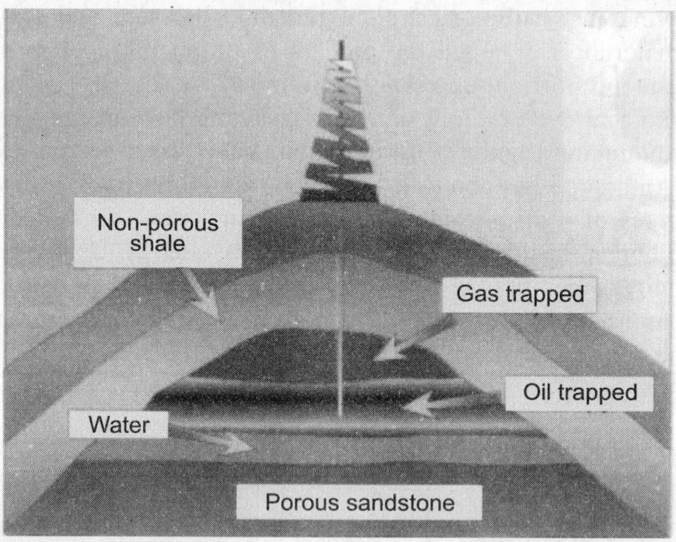

Fig. 6.5: Presence of fuel into the earth. Natural gas into the earth.

Hydrocarbons are simple molecules made up of carbon and hydrogen atoms joined together in chains or rings. These molecules, being light and mobile, migrate upwards through the rocks, but eventually become trapped beneath the impermeable rock structures in the earth's crust. Natural gas is usually found underground near an oil source. It is a mixture of light hydrocarbons including methane, ethane, propane, butane, and pentane. Other compounds found in natural gas include carbon dioxide, helium, hydrogen sulphide, and nitrogen. It is found around the world, but the largest reserves are in the former Soviet Union and the Middle East. This gas is lighter than air and is highly flammable, made up mainly of a gas called methane. Methane is a simple chemical compound that is made up of carbon and hydrogen atoms. Natural gas is usually colourless and odourless. Before this natural gas is sent into pipelines and storage tanks, it is mixed with a chemical that gives it a strong odour, almost like rotten eggs. The odour makes it easy to detect. Natural gas is the cleanest burning fossil fuel. When it is burnt, it gives off less carbon dioxide than oil or coal, virtually no sulphur dioxide, however it produces only small amounts of nitrous oxides. Natural gas is mostly composed of methane and other light hydrocarbons. Both the carbon and the hydrogen in methane combine with oxygen when natural gas is burnt, giving off heat. Coal and oil contain proportionally more carbon than natural gas, therefore giving off more carbon dioxide per unit of energy produced. Natural gas gives off 50% of the carbon dioxide released by coal and 25% less carbon dioxide than oil and produces same amount of energy. Carbon dioxide is the most important greenhouse gas contributing to global warming.

To find oil and natural gas, companies drill through the earth to the deposits deep below the surface. The oil and natural gas are then pumped from below the ground by oil rigs. They then usually travel through pipelines. At oil refineries, crude oil is split into various types of products by heating the thick black oil. The products include gasoline, diesel fuel, aviation fuel, home heating oil, oil for ships, and oil to burn in power plants to make electricity. Oil is used for transportation cars, airplanes, trucks, buses, and motorcycles. Oil is stored in large tanks until it is sent to various places to be used. Oil is used in many products — fertilizers for farms, clothes, toothbrush, plastic bottles and plastic pens. There are thousands of other products that come from oil. Almost all plastic comes originally from oil. Oil is transported in huge pipelines and tanker ships to places where it is made into other products.

The origin of oil industry in India can be traced back to the last part of the 19th century when petroleum was discovered in Digboi in north-east India. Thereafter large numbers of oil fields have been discovered both inland and off-shore. This has led to the setting up of refineries to process the oil and gas for use in various sectors.

6.4 MINERALS

A mineral is a naturally occurring solid formed through geological processes that has a characteristic chemical composition, a highly ordered atomic structure, and specific physical properties. Rock, by comparison, is an aggregate of minerals and/or mineraloids, and need not have a specific chemical composition. Minerals range in composition from pure elements and simple salts to very complex silicates with thousands of known forms. The study of minerals is called mineralogy.

India's mineral resources are sufficiently rich and varied to provide the country with a strong industrial base. The country is particularly rich in the metallic minerals, the ferrous group, such as iron ores, manganese, chromite and titanium. It has the world's largest reserves in mica and bauxite. The situation is also satisfactory in coal, fluorides, limestone, dolomite and gypsum. But the reserves of petroleum and some non-ferrous metallic minerals, especially copper, lead, zinc, tin, graphite are inadequate. Country fulfils the internal demands for these metallic minerals by importing them from other countries.

Table 6.1 shows the important mineral reserves and major minerals producing countries in the world.

Table 6.1: The status of minerals in the entire world

Mineral	Uses	World Reserves Major Producing Resources (Metric Tons)	Countries
Bauxite	Ore of aluminum	21,559,000 2	Australia, Jamaica, Brazil
Chromium	Alloys, electroplating	418,900	India, South Africa, Turkey
Copper	Alloys, electric wires	3,21,000	Chile, USA, Canada
Gold	jewellery	42	South Africa, USA, Australia
Iron ore	Iron and steel	64,648,000	Brazil, Australia, China, Canada and Venezuela
Lead	Solder, pipes	70,440	USA, Mexico, Canada
Manganese	Iron and steel	812,800	South Africa, Gabon, Australia and France
Nickel	Stainless steel	48,660	Canada, Norway and Dominican Republic.
Silver	Jewellery	780	Mexico, USA, Peru, Canada
Tin	Tin cans, alloys	5,930	China, Brazil, Indonesia
Zinc	Iron and steel	143,910	Canada, Australia, China, Peru, Mexico and Spain

In India, mineral resources are distributed unevenly. Most of the minerals are found in the ancient crystalline rocks of the Deccan and Chotanagpur plateau. Some minerals are found in the Himalayan region, although they are difficult to exploit. Besides mineral deposits, there are about 300 hot springs in ten geothermal fields, whose heat flow values vary between 83–650 MW/m^2, with thermal gradients between 70–100°C/km. The important geothermal fields are in Puga (Ladakh), Tattapani (M.P.), Jalgaon (Maharashtra) and Manikaran (H.P.), and are significant sources of underground energy. Minerals can be divided into two broad groups—metallic and non-metallic minerals. Metallic minerals can be further divided into ferrous and non-ferrous minerals (Table 6.2).

6.4.1 Metallic Minerals

6.4.1.1 *Ferrous minerals*

Ferrous minerals are those which contain iron in substantial quantity. These include iron, manganese, chromite, pyrite etc. These minerals provide a strong base for the development of metallurgical industries, particularly iron, steel and alloys.

Table 6.2: Position of minerals in India

Grouping	Adequate	Deficit	Abundance	Scarce
Fuel minerals		Coal, petroleum, crude oil	Coal	
Ferrous minerals	Chromite	Chromite, manganese ore	Iron ore	Nickel, tungsten, cobalt, vanadium
Non ferrous minerals	Zinc	Bauxite	Bauxite, copper, lead	Antimony, gold, platinum
Industrial minerals	Corundum, graphite vermiculite	Apatite, rock phosphate, asbestos Fluorite	Barites, ball clay Dolomite	Borax, potash
			Gypsum	
			Limestone	
			Mica	
			Quartzite	
			Sillimanite	
			Zircon	
Other minerals			Granite Marble	Diamond Emerald Ruby

(a) **Iron ore**: India is one of the few countries in the world, which is endowed with vast reserves of good quality iron ore. Most of the iron ore found in the country is of three types: Haematite, Magnetite and Limonite. Haematite ore contains 68% of iron, is red in colour, and is known as 'red ore'. Magnetite ore contains up to 60% of iron, is dark brown in colour and is known as 'black ore'. Limonite ore contains up to 30–50% of iron and it is yellow in colour. Iron ore deposits are found practically in every state of India. However, 96% of the total reserves are in Orissa, Madhya Pradesh, Karnataka and Goa. Orissa and Bihar together possess 50% of india's reserves of high grade iron ore. Other states like Madhya Pradesh contributes about 25% of country' total iron ore and about 20–25% of country's production of iron ore. In this way, India contributes 20% of the world's total reserves in iron, and 7-85% of total world trade from Haldia, Paradeep, Mangalore and Visakhapatnam ports.

(b) **Manganese ore**: Manganese ore forms an important ingredient in the manufacture of iron and steel. It is also used in the manufacturing of dry batteries, in photography and leather industries. India ranks third in the production of manganese ore in the world next to Russia and South Africa. The important areas of production are Orissa, Madhya Pradesh, Maharashtra, Karnataka and Andhra Pradesh. Out of these, Orissa contributes about 37% of manganese production of the total production of the country (Fig. 6.6).

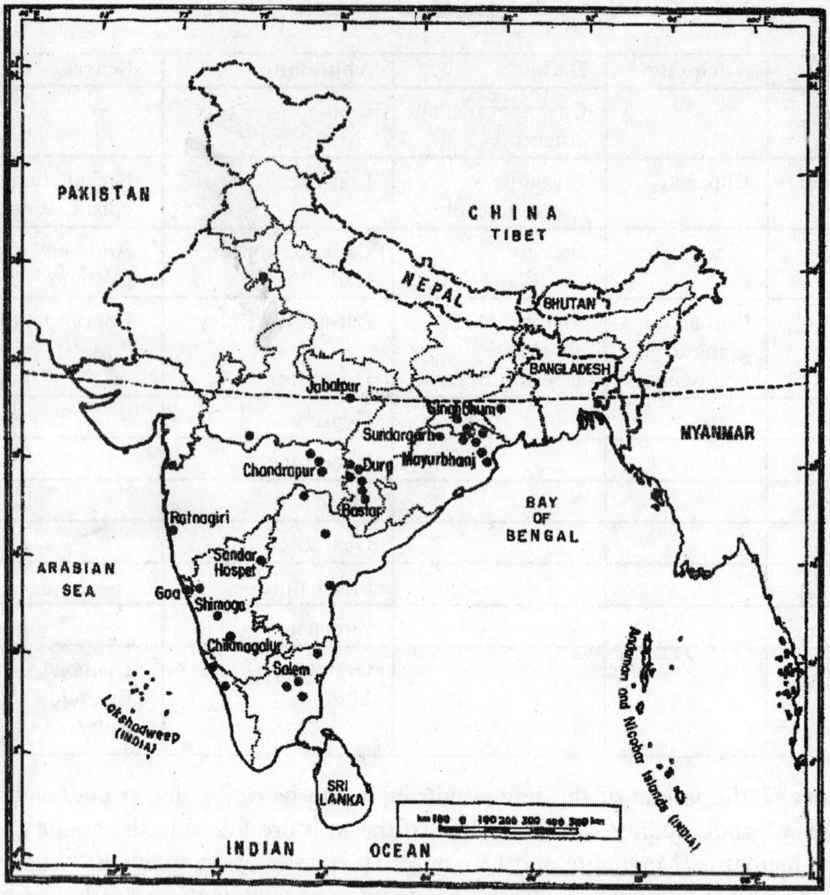

Fig. 6.6: The resources of manganese in India.

6.4.1.2 *Non-ferrous metallic minerals*

Non-ferrous minerals are those which do not contain iron. These include gold (Fig. 6.8), silver, copper, tin, lead and zinc. These metallic minerals are highly important in day-to-day life. However, India is very poor and deficient in all of these minerals.

Bauxite: Bauxite is a non-ferrous mineral. It is the ore from which aluminium metal is produced. India's reserves of bauxite are sufficient to keep the country self-reliant. Aluminium extracted from the ore is used in making areoplanes, electrical appliances and goods, household fittings etc. Bauxite is also used for manufacturing of white cement and certain chemicals. Bauxite has a wide occurrence in the country. Major reserves occur in Bihar, Maharashtra, Madhya Pradesh, Gujarat, Karnataka, Tamil Nadu and Uttar Pradesh. Bihar accounts for 13% of India's total reserves and 37% of the country's total production. The important deposits are located in Palamau, Ranchi, Munger and Lohardaga districts. Maharashtra accounts for a relatively small production of the country, 18% of the total, but possesses the second largest bauxite reserves consisting of 22% of the country's total reserves (Fig. 6.7).

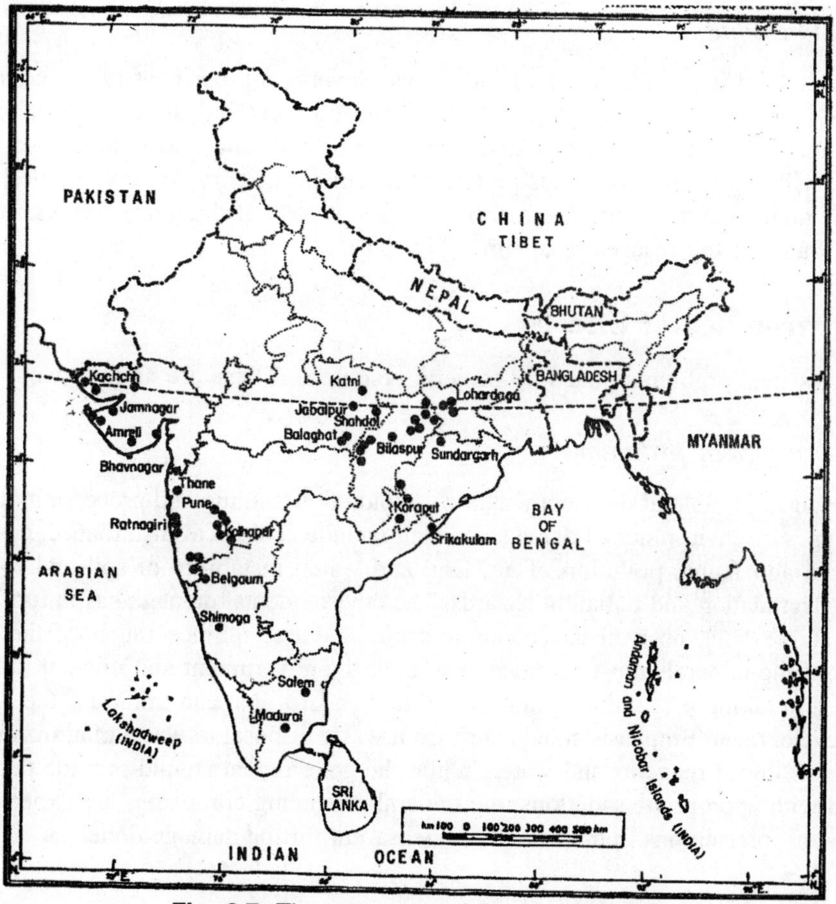

Fig. 6.7: The resources of Bauxite in India.

6.4.2 Non-metallic Minerals

A large number of non-metallic minerals are found in India, but only a few of these are commercially important. These are limestone, dolomite, mica, kyanite, sillimanite, gypsum and phosphate. These are used in industries like cement, fertilizers, and electrical goods.

6.4.2.1 *Mica*

India is the leading producer of mica. It was one of the indispensable minerals used in electrical industry till recently. Although mica is widely distributed, but workable deposits occur in three principal belts—Bihar, Andhra Pradesh and Rajasthan. Bihar produces more than 80% of mica, and Rajasthan is the third largest mica producing state. Besides these three states, some mica deposits occur in Kerala, Tamil Nadu, and Madhya Pradesh.

6.4.2.2 *Limestone*

Limestone is used in a wide range of industries. Seventy-six per cent of the country's total consumption is in cement industry, 16% in iron and steel industry, and 4% in chemical industry, and the last 4% is in sugar, paper, fertilizers and ferromanganese industries. Limestone with high silica content is preferred in cement industry. Madhya Pradesh possesses 36% of the total reserves. Karnataka contributes about 10% of the total reserves. Uttarakhand also contributes in the reservoirs of lime (Fig. 6.8).

6.4.3 Environmental Effects

There are various problems posed by mineral resources. These are as follows:

6.4.3.1 *Ecological problem*

The three stages of mineral development, viz., exploration, mining and processing, have caused different types of environmental damages, which include ecological disturbance, destruction of natural flora and fauna, pollution of air, land and water, instability of soil and rock masses, landscape degradation and radiation hazards. The environmental damage has in turn resulted in waste of arable land, as well as economic crops and trees. Since much of the damage is inevitable, if the minerals must be developed, both the government and mineral industry must be involved in taking precautionary and remedial measures that can minimize the ill-effects of mineral development. Emphasis should shift from waste disposal to waste minimization through sorting, recycling of reagents and water. While the government should provide the regulatory legislation with appropriate sanctions, the mineral-producing companies are expected to carry out mandatory precautions, remedies or compensation for the damage done.

6.4.3.2 *Pollution*

Many minerals producing areas lead to air and water pollution in the surrounding regions, which in turn leads to various health hazards.

6.4.3.3 *Social problems*

New mineral sites lead to displacement of local people. As many tribal areas are rich in minerals, the tribal people are most affected through mineral mining activity. Industrialization of such areas has badly shattered their economy, value and lifestyle.

6.4.3.4 *Depletion of minerals*

Due to excessive exploitation, many minerals are going to be depleted in near future. So it calls for conservation and judicious utilization.

6.4.4 Conservation of Mineral Resources

Scientific methods should be followed for conservation of minerals.

(1) During extraction (mining), there should be minimum waste.
(2) A judicious blend should be maintained between high-grade ores and sub-marginal or low-grade ores.
(3) Low grade ores should be subjected to the processes of beneficiation, if amenable.
(4) Ecological health of terrains subjected to mining activities must be maintained at all costs.

Fig. 6.8: The Khadia mine in Uttarakhand.

Fig. 6.9: The gold mine at Colar district.

6.5 DROUGHT

In practically all the developing countries, agriculture depends on monsoon and climatic conditions. As a result, these countries are drought prone. The Asian countries have been the largest victims of drought. Drought and flood alone are responsible for the largest number of deaths (over 53%) causing permanent damage to property and ecology. Scientists have tried to define the drought as: "*A condition of moisture deficit sufficient to have an adverse effect on vegetation, animal and man over sizeable areas*". However, scientists have identified six types of drought, namely: meteorological, climatological, atmospheric, agriculture, hydrological and water management.

In India, about 142 million hectares is net sown area. Out of this about 68% of this area is reported to be vulnerable to drought conditions, and about 50% of the drought-prone area is classified as severe, where the frequency of drought is regular. The crisis brought about directly hits the poorest and the most deprived sections of our society, thereby destroying their life, economy, infrastructure and environment.

The identification of drought is based on the precipitation, and areas can be distinguished by the rainfall. Areas that precipitate below 375 mm indicate extreme aridity, between 375 mm to 750 mm, it is semi-arid zone, between 750 mm and 1125 mm indicates a dry sub-humid zone, and the areas which receive above 1125 mm annual rainfall do not experience crop failure.

6.5.1 Prevention of Drought

Drought is a natural disaster which can be prevented through continuous monitoring of the factors, which are responsible for it. Drought prevention measures include the following three activities.

6.5.1.1 *Rainfall monitoring*

The Indian Meteorological Department (IMD), responsible for rainfall monitoring in the country, has meteorological observatories at each district headquarters and observes the weather information on a daily basis.

6.5.1.2 *Hydrological Monitoring*

Everyday, water levels in all the medium and major reservoirs are monitored. The information is carefully analyzed by the concerned state government department to make policy decisions on the amount of water to be made available for irrigation and drinking.

6.5.1.3 *Drought relief*

The state governments are primarily responsible for the management of all natural disasters, and district level administration is responsible for the operational management. The relief measures include:

- **Food security**: The procurement, storage and distribution of food grains from the food surplus areas to the deficit areas, and the distribution of essentials through the public distribution system.
- **Employment generation**: The preparation of employment generation schemes and their technical evaluation for providing jobs to the drought affected population.
- **Contingency crop plan**: An alternate crop plan with sufficient infrastructure available to stabilize farm level production.

Space technology offers immense potential for early warning and addressing of most of the issues related to drought management on a near real time basis. India has developed a space-based strategy for all the three phases of drought. Plans are afoot for a comprehensive assessment of drought by accounting for variations in rainfall, surface water availability and soil moisture, in addition to vegetation condition and communication through NICNET, for two-way data transfer to districts.

EXERCISE

QUESTIONS

1. What are the main sources of energy?
2. What are the advantages and disadvantages of renewable and non-renewable sources of energy?
3. Explain the role of winds in generation of electricity. Write the advantages and disadvantages of wind power.
4. Describe the types of minerals.
5. Define the natural disaster with example of drought and explain its prevention and control.

OBJECTIVE TYPE QUESTIONS

1. The common energy source in Indian villages is:
   (a) electricity (b) coal
   (c) sun (d) wood and animal dung
2. Fossil fuels and metallic minerals are:
   (a) renewable resources (b) non-renewable resources
   (c) inexhaustible resources (d) none of these
3. Both power and manure are provided by:
   (a) nuclear plants (b) biogas plants
   (c) inexhaustible resources (d) hydroelectric plants
4. Identify the non-renewable sources of energy:
   (a) coal (b) wind power
   (c) fuel cells (d) wave power
5. Which of the following is a disadvantage of most of the renewable energy sources?
   (a) Highly polluting (b) High running cost
   (c) High waste disposal cost (d) Unreliable supply
6. Drought is a:
   (a) natural disaster (b) man-made calamity
   (c) none of these (d) both

7. Biodiesel is produced from:
(a) vegetable oils (b) animal fats
(c) both (d) animal fats only

8. Bauxite is a:
(a) non-ferrous mineral (b) ferrous mineral
(c) both (d) none

9. Ferrous minerals contain:
(a) iron (b) lead
(c) both (d) none

10. Geothermal power is extracted from:
(a) stored heat in the earth (b) stored heat in the soil
(c) stored heat in the rocks (d) all

11. The first generation biofuels are derived from:
(a) vegetable fats (b) starch
(c) both (d) animal fats

12. The second generation biofuels are derived from:
(a) waste biomass (b) vegetable fats
(c) both (d) none of them

13. The third generation biofuels are derived from:
(a) algae (b) animal fats
(c) waste biomass (d) all

14. The fourth generation biofuels are derived from:
(a) microorganisms (b) natural oil
(c) animals (d) all

15. Renewable energy sources:
(a) regenerate continuously (b) regenerate in short period
(c) regenerate in long period (d) none

16. The study of minerals is called:
(a) mineralogy (b) mineology
(c) meteorology (d) none

17. Hydroelectric energy is generated by:
(a) kinetic energy of water (b) kinetic energy of wind
(c) kinetic energy of aquatic life (d) none

18. All types of coals have a harmful contain
(a) sulphur (b) nitrogen
(c) hydrocarbon (d) all

19. Types of coal:
(a) lignite (b) bituminous
(c) anthracite (d) all

20. Non-renewable energy sources accumulated in nature:
    (a) over a long time (b) within 10 years
    (c) none (d) both

ANSWERS

1. (d) 2. (b) 3. (b) 4. (a) 5. (c) 6. (a) 7. (a) 8. (a) 9. (a) 10. (a)
11. (a) 12. (a) 13. (a) 14. (a) 15. (a) 16. (a) 17. (a) 18. (c) 19. (d) 20. (a)

7

ENVIRONMENTAL LAWS

7.1 INTRODUCTION

Environmental laws come under the federal law which was established to protect the nature / environment in which we live. This includes ensuring the health and safety of the flora and fauna in the United States. In 1970, the federal government established the Environmental Protection Agency (EPA) to enforce environmental law established by the government. This agency was created in response to public demand for cleaner air, water, and land.

Also in 1970, the **National Environment Policy Act** (NEPA) was passed by the legislature. This was the first environmental law developed in the United States. It set forth a broad national environmental policy that would prevent or eliminate environmental damage, enrich our understanding of the environment, ecology, and natural resources, and create a Council on Environmental Quality.

In India, Environment Protection Act was established in 1986; however in the Constitution of India it is clearly stated that it is the duty of the state to 'protect and improve the environment and safeguard the forests and wildlife of the country'. It imposes a duty on every citizen 'to protect and improve the natural environment including forests, lakes, rivers, and wildlife'. Reference to the environment has also been made in the Directive Principles of State Policy as well as the Fundamental Rights. The Department of Environment was established in India in 1980 to ensure a healthy environment for the country. This later became the Ministry of Environment and Forests in 1985.

The EPA (Environment Protection Act), 1986, came into force soon after the Bhopal Gas Tragedy (December 1984), and is considered an umbrella legislation as it fills many gaps in the existing laws. Thereafter, a large number of laws came into existence as the problems began arising, for example, Handling and Management of Hazardous Waste Rules in 1989. Different Acts and their embedment have been mentioned below. Following is a list of the environmental legislations that have come into effect.

7.2 ENVIRONMENT PROTECTION ACT, 1986

The Environment Protection Act authorizes the central government to protect and improve environmental quality, control and reduce pollution from all sources, and prohibit or restrict the setting and /or operation of any industrial facility on environmental grounds. Embedment 1989, the objective of Hazardous Waste (Management and Handling) Rules is to control the generation, collection, treatment, import, storage, and handling of hazardous waste.

Environment Protection Act also includes the Public Liability Insurance Act, and Rules and Amendment, 1992 was drawn up to provide public liability insurance for the purpose of providing immediate relief to the persons affected by accident while handling any hazardous substance.

Environment Protection Act, 1998, described the biomedical waste (Management and Handling) Rules a legal binding on the healthcare institutions to streamline the process of proper handling of hospital waste such as segregation, disposal, collection, and treatment. The Municipal Solid Wastes (Management and Handling) Rules, 2000 apply to every municipal authority responsible for the collection, segregation, storage, transportation, processing, and disposal of municipal solid wastes, as well as the Ozone Depleting Substances (Regulation and Control) Rules have been laid down for the regulation of production and consumption of ozone depleting substances.

In 2002, under Environment Protection Act, the Noise Pollution (Regulation and Control) (Amendment) Rules laid down such terms and conditions as are necessary to reduce noise pollution, permit the use of loud speakers or public address systems during night hours (between 10:00 p.m. to 12:00 midnight) on or during any cultural or religious festive occasion, and the Biological Diversity Act (2002) to provide for the conservation of biological diversity, sustainable use of its components and equitable sharing of the benefits arising out of the use of biological resources and knowledge associated with it.

7.3 WILDLIFE PROTECTION ACT, 1972

The Wildlife Protection Act, 1972, Rules 1973 and Amendment 1991 provides for the protection of birds and animals and for all matters that are connected to it whether it be their habitat or the waterhole or the forests that sustain them. The Forest Conservation Act, 1980 and Rules provide for the protection and conservation of the forests. In accordance with Wildlife Protection Amendment Act, 2002, "no alternation of boundaries/National Park/ Sanctuary shall be made by the State Govt. except on recommendation of the National Board for Wildlife (NBWL)". Whenever, any part of Wildlife Sanctuary/National Park is getting affected by a hydro project, the forest proposal in respect of such project is entertained by MoEF, GOI only after the permission of de-reservation/denotification of Wildlife Sanctuary/ National Park has been accorded. After recommendation of Standing Committee of NBWL, proposal for de-reservation/ denotification is ratified by Hon'ble Supreme Court.

Wildlife Protection Act, 1972

Section	Content
Section 9 and 11	Those wild animals which are included in schedule I, II and III cannot be hunted. Killing of a dangerous wild animal in self-defence would not be considered an offence. Permission may be granted for hunting wild animals for special purposes such as education, scientific research and scientific management.
Section 17 A, 17 B, 17 C and 17 D	Picking, uprooting, etc., of specified plants is strictly prohibited. Permission may be granted for special purposes such as education, scientific research, etc.

Section 18 and 27	State government can declare area as a sanctuary because of its ecological, faunal, floral, geomorphological, natural, or zoological significance. Other than specified persons, no one can enter a sanctuary. Destruction, fire, entry with weapons and use of hazardous substance in a sanctuary are prohibited.
Section 38	Empowers Central Government to declare an area as a sanctuary or national park. There is a provision of constituting Central Zoo Authority by Central Government.

7.4 WATER PREVENTION AND CONTROL OF POLLUTION ACT, 1974

The objectives of the Water Prevention and Control of Pollution Act are to provide for the prevention and control of water pollution and the maintenance or restoration of the wholesomeness of water for the establishment, with a view to carry out the purposes aforesaid, of Boards for the prevention and control of water pollution, for conferring on and assigning to such Boards powers and functions relating thereto and for matters connected therewith. The CPCB (Central Pollution Control Board) was constituted under this Act.

Water Prevention and Control of Pollution Act, 1974

Section	Content
Sections 3,4,8 and 13	In this Act, there is provision of constitution of central and state Boards with a full time chairman and 16 members. The chairman of state pollution control board could be part time as well.
Sections 21 and 23	Person empowered by the State Board has the power to take samples of effluents and procedure to be followed in connection therewith power of entry and inspection.
Sections 24 and 32	Prohibition on use of streams or wells for the disposal of polluting matter. Emergency measures can be taken in case of pollution of stream or well. The State Board is empowered for such measures.
Sections 63 and 64	Powers of the Central Government and State Government to make rules

7.5 AIR PREVENTION AND CONTROL OF POLLUTION ACT, 1981

The objective of this Act is to provide for the prevention, control and abatement of air pollution, for the establishment, with a view to carry out the aforesaid purposes, of Boards, for conferring on and assigning to such Boards powers and functions relating thereto and for matters connected therewith. Decisions were taken at the United Nations Conference on the Human Environment held in Stockholm in June 1972, in which India participated to take appropriate steps for the preservation of the natural resources of the earth which, among other things, includes the preservation of the quality of air and control of air pollution. ***(sources from** TERI report, MOEF **website)***

Air Prevention and Control of Pollution Act, 1981

Section	**Content**
Section 3	Provision for constitution of Central Board for prevention and control of air pollution.
Section 5	Provision for constitution of State Boards.
Section 19	Power to declare air pollution control areas and prohibition for use of fuel, appliance, and burning material other than those approved in such areas.
Section 22	Industrial activities are not allowed to emission of air pollutants to excess the standards laid down by the State Board.
Sections 24 and 26	Person empowered by the State Board shall have the right to enter and inspect any place and take samples of air or emission and procedure to be followed in connection therewith.
Sections 40 and 41	Provision of punishment for offences by the companies and offences by the government department.
Section 47	Power of the State Government to supersede State Board in the public interest or other special circumstances.
Sections 53 and 54	Powers of the State and Central Governments to make rules.

Non-governmental organizations (NGOs) can and have played a significant role in environmental protection across the world and in India. The success of environmental programmes primarily depends upon the awareness and consciousness of the people. It is essential to sensitize the people on environmental problems so as to bring about a remarkable change in the condition of environment and related aspects.

India has a large network of NGOs, which are involved in spreading the message of sustainable development. The following is the list of some of the popular NGOs in the country.

1. **Bombay Natural History Society:** BNHS is actively involved in the study of wildlife related problems and subsequent recommendation to conserve the nature through Environment Management plan and Environment Impact Assessment studies.
2. **Centre for Environmental Education:** It conducts widespread environmental education, awareness and training programmes. It has also taken up projects related to pollution control, biodiversity conservation and sustainable development.
3. **Centre for Science and Environment**: CSE is the India's first environmental NGO to analyze and study the relationship between environment and development and create public consciousness about the need for sustainable development.
4. **Tata Energy and Research Institute:** TERI is a independent research institute which is working on research and development activity and promoting environmental technology, environmental policies and sustainable development use of natural resources.

EXERCISE

QUESTIONS

1. What is the role of NGOs?
2. Outline the chief provisions of Environmental Protection Act, 1986.
3. Discuss the Wildlife Protection Act, 1972.
4. Discuss the Environmental Acts in brief.

OBJECTIVE TYPE QUESTIONS

1. The objectives of Wildlife Protection Act, 1972 are:
   (a) to preserve biodiversity
   (b) to maintain essential ecological and life support system
   (c) protection and conservation of wildlife
   (d) all of them
2. The Forest Act was enacted in:
   (a) 1986 (b) 1974
   (c) 1980 (d) 1972
3. Environmental education should be imparted only at:
   (a) primary school (b) secondary stages
   (c) college stage (d) all stages
4. The Environment Protection Act was enacted in:
   (a) 1987 (b) 1989
   (c) 1986 (d) 1985
5. The Air Prevention and Control of Pollution Act was enacted in:
   (a) 1981 (b) 1982
   (c) 1983 (d) 1984
6. The first major environmental protection act to be promulgated in India was:
   (a) Water Act (b) Air Act
   (c) Environmental Act (d) Noise Pollution Rules
7. The objective of environmental education is to:
   (a) raise consciousness about environmental conditions
   (b) teach environmentally appropriate behaviour
   (c) both
   (d) none
8. Environmental education as a subject was announced by the order of:
   (a) High Court, Delhi (b) Supreme Court
   (c) Environmental Act 1986 (d) None of these

9. Biodiversity Act was enacted in:
   (a) 2001 (b) 2002
   (c) 2003 (d) 2004
10. How can we protect our natural resources?
    (a) Through education (b) Through Laws and policy
    (c) To promote our indigenous knowledge (d) All of these

ANSWERS

1. (c) 2. (c) 3. (d) 4. (c) 5. (a) 6. (c) 7. (c) 8. (b) 9. (b) 10. (d)

Question Paper of Environmental Studies in UTTARAKHAND TECHNICAL UNIVERSITY

TES-101 **Paper ID: 0099**

B. Tech.

(SEM.1) (ODD SEM.) EXAMINATION, 2009-10

ENVIRONMENTAL STUDIES

Time: 2.00 hr. **Total Marks: 50**

1. The objective of environmental education is to:
   (a) raise consciousness about environmental conditions
   (b) teach environmentally appropriate behaviour
   (c) create an environmental ethic that forests awareness about ecological interdependence
   (d) all of the above
2. Which of the following is not influenced by human activities?
   (a) Depletion of groundwater
   (b) Destruction of mangroves and wetlands
   (c) Increased extinction rates of species
   (d) None of the above
3. Which of the following statement about environmental education is false?
   (a) Environmental education is evolving to be education for sustainable and ethical development
   (b) Environmental education will prepare the next generation to plan appropriate strategies for addressing developmental environmental issues
   (c) Environmental education does not advocate a particular viewpoint or course of action
   (d) Environmental education is essential only for the younger generation
4. Which of the following conceptual spheres of the environment has the least storage capacity for matter?
   (a) Atmosphere (b) Lithosphere
   (c) Hydrosphere (d) Biosphere
5. Biosphere is:
   (a) the solid shell of organic matter on the surface of the earth comprising all the living things.
   (b) the thin shell of organic matter on the surface of the earth comprising all the living things.

(c) the sphere which occupies the maximum volume of all the spheres.
(d) all of the above

6. Global atmosphere temperature is likely to be increased due to:
(a) burning of fossil fuel
(b) water pollution
(c) soil erosion
(d) none of the above

7. Which of the following statements about forest is not correct?
(a) Forest reduces soil erosion
(b) Provides recreational opportunities
(c) Provides a source of economic development
(d) None of the above

8. Which of the following is not true about deforestation?
(a) Population explosion is not the reason of deforestation
(b) Clearing of forest for agriculture causes deforestation
(c) Deforestation is taking place only in developing countries
(d) Cash crop economy of the third world is a cause of deforestation

9. Selling of carbon by the developing industries is known as:
(a) carbon sequestration
(b) carbon dating
(c) carbon credit
(d) none of the above

10. Major purpose of most of the dams around the world is:
(a) tourist development
(b) irrigation
(c) drinking water supply
(d) flood control

11. Which of the following is not a major solution for tackling the water crisis?
(a) Drilling large number of deep bore wells
(b) Population growth control
(c) Water conservation in irrigation
(d) Water pollution control

12. Which of the following is the most eco-friendly agriculture?
(a) Use of chemical fertilizers and insecticides
(b) Use of insecticides and organic fertilizers
(c) Use of organic fertilizers and alternate methods for insect controlled
(d) Use of chemical fertilizers and alternate methods for insect control

13. Which of the following could be the most important socio-economic impact of dams?
(a) Loss of biodiversity
(b) Poorly managed involuntary displacement and loss of livelihood
(c) Waterlogging
(d) Loss of forests and wildlife habitat

14. Identify the non-renewable sources of energy from the following.
(a) Coal
(b) Fuel cells
(c) Wind power
(d) Wave power

15. Which of the following forms of land degradation is more prevalent in India?
    (a) Desertification (b) Soil erosion
    (c) Landslide (d) Soil subsidence
16. Which of the following is a possible producer in an ecosystem?
    (a) Plants and some bacteria (b) Animals
    (c) Human beings (d) Fish
17. Which of the following statement is false?
    (a) Inorganic nutrients are recycled in an ecosystem
    (b) Energy flows through the ecosystem in the form of carbon-carbon bonds
    (c) Energy is recycled in an ecosystem
    (d) All
18. A food web consists of:
    (a) a portion of a food chain
    (b) Producers, consumers and decomposers
    (c) interlocking food chains
    (d) a set of odd similar consumers
19. A tropic level refers to:
    (a) area in the tropics
    (b) an organism's position in an ecosystem
    (c) an organism's position in a food chain
    (d) an organism's position in a biome
20. Tendency of pollutants to become concentrated in successive trophic levels is known as:
    (a) bioremediation (b) biomagnification
    (c) biopiracy (d) biorhythm
21. Primary component in the biochemical cycle is:
    (a) human beings (b) large animals
    (c) water (d) plants and microbes
22. The largest reservoirs of nitrogen on our planet is:
    (a) oceans (b) atmosphere
    (c) biosphere (d) fossil fuel
23. Eutrophication is:
    (a) an improved water quality status of lakes
    (b) the result of accumulation of plant nutrients in waterbodies
    (c) a process in the carbon cycle
    (d) nutrients
24. Genetic variation between distinct population of the same species is known as:
    (a) species diversity (b) ecosystem diversity
    (c) genetic diversity (d) biodiversity

25. Which is not associated with the structure of forest?
(a) Canopy
(b) Under story
(c) Forest floor
(d) First floor

26. In an aquatic ecosystem, phytoplankton can be considered:
(a) consumers
(b) producers
(c) saprotrophic organism
(d) nutrients

27. Which of the following is not a world heritage site?
(a) Manas Wildlife Sanctuary
(b) Nanda Devi National Park
(c) Kaziranga National Park
(d) Periyar National Park

28. Which of the following is an endemic species of India?
(a) Asian elephant
(b) Lion tailed macaque
(c) Whales
(d) Panda

29. Which of the following is a biodiversity hot spot in India?
(a) Gulf of Mannar
(b) Western Ghats
(c) Panchmarhi
(d) Sunderbans

30. Which of the following is an *in situ* conservation area?
(a) National parks
(b) Captive breeding centres
(c) Botanical gardens
(d) Zoos

31. Which of the following is a cause of loss of biodiversity?
(a) Habitat degradation and loss
(b) Invasion of non-native species
(c) Pollution
(d) All of these

32. Which of the following is an air pollutant?
(a) Nitrogen
(b) Carbon monoxide
(c) Carbon dioxide
(d) Oxygen

33. Which of the following statements about carbon is true?
(a) It is a result of incomplete combustion of fossil fuel
(b) It is harmless to human beings
(c) It is a foul smelling gas
(d) All of the above

34. Which of the following is a secondary air pollutant?
(a) Carbon monoxide
(b) Sulphur dioxide
(c) Ozone
(d) Carbon dioxide

35. Smog is:
(a) a natural phenomenon
(b) a combination of smoke and fog
(c) colourless
(d) all of the above

36. Air pollution from automobiles can be controlled by fitting:
(a) electrostatic precipitator
(b) wet collector
(c) catalytic converter
(d) all of the above

37. Which is the permissible range of pH for drinking water as per the Indian standards?
(a) 6–9 (b) 6.5–7.5
(c) 6–8.5 (d) 7.5–9.5
38. Excess of iron in water is likely to cause:
(a) colour (b) taste
(c) hardness (d) all of the above
39. Which of the following is non-point source of water pollution?
(a) Factories (b) Sewage treatment plants
(c) Urban and suburban lands (d) All of the above
40. Domestic sewage is:
(a) wastewater generated from kitchen
(b) wastewater generated from residential area
(c) water entering a sewer as a result of rainfall
(d) wastewater from industries
41. In Indian conditions, which of the following sewerage systems is ideal?
(a) Combined system
(b) Separate system
(c) Partially combined or partially separate system
(d) All of the above
42. BOD is:
(a) biological oxygen degradation (b) biological oxygen development
(c) biochemical oxygen demand (d) all of the above
43. Noise is:
(a) loud sound (b) unwanted sound
(c) constant sound (d) sound of high frequency
44. Biomedical waste management is:
(a) incineration (b) autoclaving
(c) both (a) & (b) (d) none of the above
45. How many Biosphere Reserve Forests are present in the Uttarakhand State?
(a) Two (b) One
(c) Three (d) Six
46. Which is the Uttarakhand State bird?
(a) Himalayan Quail (b) Himalayan Peafowl
(c) Himalayan Monal (d) None of the above
47. Which is the Uttarakhand State animal?
(a) Spotted deer (b) Snow leopard
(c) Musk deer (d) None of these
48. Which of the following is not a method of water conservation?
(a) Rainwater harvesting (b) Groundwater extraction
(c) Improving irrigation efficiency (d) Avoiding water wastage

49. In India, a major cause of large scale displacement of people is:
(a) forest fires
(b) development project
(c) earthquakes
(d) war

50. The primary cause of acid rain around the world is:
(a) carbon dioxide
(b) sulphur dioxide
(c) carbon monoxide
(d) ozone

51. Ozone layer is present in:
(a) troposphere
(b) stratosphere
(c) mesosphere
(d) thermosphere

52. Global warming could effect:
(a) climate
(b) food production
(c) melting of glaciers
(d) all of the above

53. First of the major environmental protection acts to be promulgated in india was:
(a) the Water Act
(b) the Air Act
(c) the Environment Act
(d) the Noise Act

54. Full form of FSI is:
(a) forest services of India
(b) forest special insurance
(c) forest survey of India
(d) forest survey institute

55. Full form of UNDP is:
(a) unity in national development programme
(b) union nation development planning
(c) united nation development planning
(d) united nation development programme

56. How many hot spots are there in India?
(a) Three
(b) Four
(c) Two
(d) One

57. How many biosphere reserves are there in India?
(a) Twelve
(b) Eleven
(c) Sixteen
(d) Fourteen

58. One-horn Rhino is an endemic species of India:
(a) false
(b) in past
(c) true
(d) none of the above

59. Wildlife Protection Act was established in India in the year:
(a) 1974
(b) 1992
(c) 1975
(d) 1972

60. Commonwealth Government recognized four elements in disaster management:
(a) PCCP
(b) PRCP
(c) PRRP
(d) PRRC

61. *Ex situ* area of biodiversity conservation is:
(a) biosphere reserve forest (b) botanical gardens
(c) wetlands (d) Ramsar wetlands

62. Formation of algal bloom in waterbody is due to the:
(a) enrichment of plant diversity (b) enrichment of animal density
(c) enrichment of hazardous pollutants (d) enrichment of nutrients

63. The function of ecosystem is:
(a) flow of chemical nutrients (b) flow of various gases
(c) flow of water (d) none of the above

64. Full form of IUCN is:
(a) international union for contribution in nature
(b) Indian union for control of natural gases
(c) international unity in conservation of nature and natural resources
(d) international union for conservation of nature

65. Which of the following is a management option for air pollution?
(a) Regulations and standards (b) Emission changes
(c) Transport planning (d) All of the above

66. Which of the following is an example of impact of development activities on the hydrosphere?
(a) Air pollution (b) Water pollution
(c) Burning of fossil fuel (d) None of the above

67. Both power and manure is provided by:
(a) nuclear plants (b) thermal plants
(c) biogas plants (d) hydroelectric plant

68. Common energy source in Indian villages is:
(a) electricity (b) coal
(c) sun (d) wood

69. Which of the following is not a viable protection against deforestation?
(a) Reduce the consumption of forest
(b) Boycott production of companies involved in deforestation
(c) Privatization of forest land
(d) Environmental education

70. Which of the following is a disadvantage of most of the renewable energy sources?
(a) Highly polluting (b) High waste disposal cost
(c) Unreliable supply (d) High running cost

71. The food chain in which microorganisms break down the dead producers is called:
(a) consumer food chain (b) predator food chain
(c) parasitic food chain (d) detritus food chain

72. Which ecological pyramid is always upright?
(a) Pyramid of energy (b) Pyramid of biomass
(c) Pyramid of numbers (d) Pyramid of biomass and numbers
73. Bhopal tragedy was due to the leakage of:
(a) methyl sodium (b) sodium isothiocyanate
(c) methyle isocyanate (d) potassium isothiocyanate
74. The year of 'Bhopal Gas Tragedy' is:
(a) 1994 (b) 1985
(c) 1986 (d) 1984
75. The disease caused by eating fish inhabiting mercury contaminated water is:
(a) Bright's disease (b) Itai-itai disease
(c) Minamata disease (d) None of the above
76. Which of the following is not a greenhouse gas?
(a) Oxygen (b) Methane
(c) CFCs (d) Carbon dioxide
77. Biological Diversity Conservation Act in India was established in the year:
(a) 2004 (b) 2002
(c) 2006 (d) 2001
78. ISO 14000 and 14001 stand for:
(a) environmental management statement
(b) environmental conservation and control
(c) environmental management systems
(d) environment education programme
79. BNHS is:
(a) Bureau of Natural and Hazardous System
(b) Bombay Natural History Society
(c) Bureau of Nature Habitat Structure
(d) Bombay Natural Historical Society
80. MoEF is known as:
(a) monitoring of environment and forest
(b) ministry of environment and forest
(c) municipality of environmental forum
(d) none of the above
81. The environment which has been modified by human activities is called:
(a) natural environment (b) anthropogenic environment
(c) urban environment (d) modern environment
82. The part of the earth and its atmosphere in which organisms live is:
(a) biosphere (b) biome
(c) biomass (d) biota

83. Global atmospheric temperature is likely to increase due to:
(a) burning of fossil fuel (b) water pollution
(c) soil erosion (d) none of the above

84. Atmosphere consists of 79% nitrogen and 21% of oxygen by:
(a) volume (b) weight
(c) density (d) all of the above

85. The term ecology was introduced by:
(a) Hackle (b) Odum
(c) Tansely (d) Ramdeo Mishra

86. The ecosystem consists of:
(a) green plants and animals
(b) green plants and decomposers
(c) producers and consumers
(d) green plants, decomposers and abiotic environment

87. Which one of the following is an abiotic component of the ecosystem?
(a) Bacteria (b) Plants
(c) Humus (d) Fungi

88. Increase in fauna and decrease in flora would be harmful due to increase in:
(a) disease (b) carbon dioxide
(c) oxygen (d) radioactive pollution

89. Tropical rain forests occur in India in:
(a) Jammu and Kashmir (b) Rajasthan
(c) Kerala and Assam (d) Nowhere

90. The largest reservoir of nitrogen on our planet is:
(a) ocean (b) atmosphere
(c) biosphere (d) fossil fuel

91. The most stable ecosystem is:
(a) forest (b) desert
(c) ocean (d) mountain

92. Production, transformation and use of energy are the major problems of:
(a) agriculture (b) transportation
(c) mining (d) sustainable development

93. Fossil fuel and mineral resources are
(a) renewable (b) inexhaustive resources
(c) non-renewable resources (d) none of the above

94. Which of the following statements about the forest is not correct?
(a) Forest reduces soil erosion (b) Provides recreational opportunities
(c) Provides economic development (d) None of the above

95. Deforestation generally decreases:
(a) rainfall (b) soil erosion
(c) drought (d) global warming
96. Extensive planting of trees to increase forest cover is called:
(a) afforestation (b) agro forestry
(c) deforestation (d) social forestry
97. Chipko Movement in Karnataka is known as:
(a) Appiko movement (b) Sipko movement
(c) Assiko movement (d) None of these
98. Chipko movement was started to conserve:
(a) forest (b) desert
(c) grassland (d) mountain
99. Amrita Devi is known for:
(a) water conservation (b) soil conservation
(c) wildlife conservation (d) tree conservation
100. Which of the following is a non-conventional energy resource?
(a) CNG (c) Tidal
(b) LPG (d) Nuclear

ANSWERS

1. (d)	2. (c)	3. (c)	4. (a)	5. (d)	6. (a)	7. (b)	8. (a)	9. (a)	10. (b)
11. (a)	12. (b)	13. (b)	14. (a)	15. (a)	16. (a)	17. (b)	18. (c)	19. (b)	20. (b)
21. (d)	22. (b)	23. (b)	24. (c)	25. (d)	26. (b)	27. (a)	28. (b)	29. (b)	30. (a)
31. (a)	32. (b)	33. (b)	34. (d)	35. (b)	36. (d)	37. (b)	38. (c)	39. (c)	40. (b)
41. (c)	42. (c)	43. (b)	44. (a)	45. (b)	46. (c)	47. (c)	48. (b)	49. (b)	50. (b)
51. (b)	52. (a)	53. (a)	54. (c)	55. (d)	56. (c)	57. (c)	58. (c)	59. (d)	60. (c)
61. (b)	62. (d)	63. (a)	64. (c)	65. (d)	66. (b)	67. (c)	68. (d)	69. (c)	70. (c)
71. (d)	72. (a)	73. (c)	74. (b)	75. (c)	76. (a)	77. (b)	78. (c)	79. (b)	80. (b)
81. (b)	82. (a)	83. (a)	84. (c)	85. (a)	86. (c)	87. (c)	88. (b)	89. (c)	90. (b)
91. (c)	92. (d)	93. (c)	94. (c)	95. (a)	96. (a)	97. (a)	98. (a)	99. (d)	100. (d)

TES-101/201 **Paper ID: 0099**

B.Tech.
ENVIRONMENTAL STUDIES

Time: 2.00 hr. **Total Marks: 50**

1. Which of the following habitats is most unsuitable for primary productivity?
   (a) Meadow (b) Forest
   (c) Cave (d) Pond
2. Maximum contribution of O_2 is from:
   (a) phytoplankton (b) grassland
   (c) herb & shrubs (d) dense forest
3. Ecosystems create:
   (a) food chain (b) food web
   (c) both a & b (d) none of the above
4. In lake ecosystem pyramid of biomass is
   (a) upright (b) inverted
   (c) anything is possible (d) none of them
5. Man is:
   (a) producer (b) primary consumer
   (c) top consumer (d) primary & secondary consumer
6. Which one of the following is a top carnivore of grassland food chain?
   (a) Snake (b) Dear
   (c) Hawk (d) Frog
7. The second most important trophic level in lake is:
   (a) phytoplankton (b) zooplankton
   (c) *Rhizopus* (d) *Penicillium*
8. Which is the most stable ecosystem?
   (a) Desert (b) Ocean
   (c) Mountain (d) Forest
9. Which of the following is the nature's cleaner?
   (a) Producers (b) Consumer
   (c) Decomposers (d) Man
10. Biomass of producers within a specified area will be high in:
    (a) forest (b) grassland
    (c) lake ecosystem (d) pond ecosystem
11. Red Data Book refers to:
    (a) a book with red cover (b) a book with a list of rare/extinct species
    (c) a book with a list of red species (d) book written by Karl Marx

12. Total biogeographical zones in India are:
(a) 5 (b) 7
(c) 12 (d) 10

13. Hot spot area refers to:
(a) hot climate area (b) endemic species area
(c) endangered species area (d) rare species area

14. Who is the Union Minister of Environment and Forestry?
(a) Mr. Jai Ram Ramesh (b) Dr. Manmohan Singh
(c) Mr. T.R. Balu (d) Mr. T.R. Sankarraman

15. Who is the Environment and Forest Minister of Uttarakhand?
(a) Mr. Bansidhar Bhagat (b) Mr. Bisan S. Chaupal
(c) Dr. Ramesh Pokhriyal (d) Mr. Madan Kausik

16. CITES refers to:
(a) Conservation for Indian Tiger and Ecosystems
(b) Conservation on International Trade in Endangered Species
(c) Convention on International Trade in Endangered Species

17. Wildlife Protection Act was passed in:
(a) 1971 (b) 1972
(c) 1978 (d) 1974

18. Project Tiger was launched in:
(a) 1965 (b) 1973
(c) 1972 (d) 1975

19. World Biodiversity day is celebrated on:
(a) 12 May (b) 23 May
(c) 24 May (d) 22 May

20. Total protected area in India is:
(a) 112 (b) 581
(c) 584 (d) 589

21. Biodiversity Act was passed in India in:
(a) 2000 (b) 2001
(c) 2002 (d) Still waiting

22. Which of the following species has become extinct?
(a) Dodo (b) Red headed vulture
(c) Eagle (d) Hawk

23. IUCN refers to:
(a) International Union for Conservation Nature
(b) International Union for Conservation Nature and Natural Resources
(c) International Union for Conserving Nature

24. UNEP refers to:
    (a) United Nations Environmental Project
    (b) United Nations Environment Programme
    (c) United Nations Environment Protection
    (d) United Nations Environment Protection Act
25. Which one of the following is not a wildlife sanctuary?
    (a) Motichur (b) Binsar
    (c) Naina (d) Nanda Devi
26. How many hot spots area have been introduced?
    (a) 12 (b) 24
    (c) 34 (d) 39
27. Indian biodiversity stands at which place at global level:
    (a) 1 (b) 12
    (c) 10 (d) 20
28. National Parks come under:
    (a) *In situ* (b) *Ex situ*
    (c) zoo (d) biosphere reserve
29. Biosphere refers to
    (a) conserve the wildlife (b) conserve the birds
    (c) conserve the vegetation (d) conserve the ecosystem
30. How many wetlands have been introduced in India?
    (a) 12 (b) 15
    (c) 25 (d) 10
31. Wetland sites are also known as:
    (a) Rem sites (b) Ramsar sites
    (c) Ramsor sites (d) Ramsey sites
32. How many biogeographical zones are there in the world?
    (a) 10 (b) 12
    (c) 14 (d) 20
33. What is the state flower of Uttarakhand?
    (a) Lotus (b) Burans
    (c) Bram-kamal (d) Rose
34. The term biodiversity was introduced in:
    (a) 1990 (b) 1992
    (c) 1994 (d) 2001
35. How can we conserve our biodiversity?
    (a) Through awareness (b) Through indigenous knowledge
    (c) Policies and laws (d) Through education

36. Algal bloom is caused by:
(a) diatoms
(b) green algae
(c) blue-green algae
(d) all algae
37. Smog is:
(a) colourless
(b) natural phenomenon
(c) smog and fog
(d) all of these
38. Leakage of MIC in Bhopal occurred on:
(a) 3rd Dec, 1984
(b) 5th Dec, 1986
(c) 7th Dec, 1987
(d) 6th Dec, 1984
39. Environmental factors are:
(a) biotic / abiotic
(b) living / non-living
(c) both
(d) wood and animal
40. Biotic factors are:
(a) plant, animals & microbes
(b) man and animal
(c) wind & water
(d) none
41. Environmental Science subject is:
(a) scientific
(b) non-scientific
(c) technical
(d) science and social subject
42. Who proposed the term ecology?
(a) E. Hackel
(b) Thomson
(c) A.G. Tansley
(d) None
43. Atmosphere is composed of:
(a) troposphere
(b) stratosphere
(c) mesosphere
(d) all
44. Environmental Protection Act was proposed in:
(a) 1986
(b) 1896
(c) 1996
(d) 1869
45. A main source of energy in environment is:
(a) coal
(b) LPG
(c) sun
(d) wind
46. Greek '*oikos*' means:
(a) plants, animals & microbes
(b) man and animal
(c) household
(d) none
47. The largest amount of usable water found on the earth is used for:
(a) industrial uses
(b) irrigation
(c) household
(d) crop irrigation
48. The gas in the atmosphere that can absorb the UV radiation:
(a) O_3
(b) SO_2
(c) CO_2
(d) none

49. GIS means:
    (a) Geographic Information System (b) Geographic Information Satellite
    (c) Both (d) Geographic Institute for Studies
50. Environmental studies is:
    (a) plants & animals (b) man & nature
    (c) wind & water (d) development with nature
51. Pedology is the branch of science in which we study about:
    (a) soil (b) rocks
    (c) water (d) animal
52. Lithosphere is related to:
    (a) land & soil (b) water & soil
    (c) wind & water (d) none
53. Hydrosphere is related to:
    (a) water & soil (b) living / non-living
    (c) water (d) all
54. Atmosphere is a part of:
    (a) gases (b) water
    (c) wind (d) none
55. The branch of science in which we study about nature and its components is:
    (a) ecology (b) environmental science
    (c) both (d) none
56. Snowfall is:
    (a) biotic factor (b) abiotic factor
    (c) both (d) none
57. Edaphic factors are:
    (a) soil (b) water
    (c) wood (d) all
58. Microbes are:
    (a) biotic factor (b) abiotic factor
    (c) edaphic factor (d) none
59. The darker zone in lakes where light penetration is negligible is called:
    (a) littoral zone (b) limnetic zone
    (c) profound zone (d) euphotic zone
60. Global warming could affect:
    (a) climate (b) food production
    (c) melting of glaciers (d) all of these
61. ISO 14000 and 14001 stand for:
    (a) Environmental management statement (b) Environment conservation and control
    (c) Environmental management system (d) Environmental Education

62. BNHS is:
(a) Bureau of natural and hazardous system (b) Bombay natural historical society
(c) Bombay natural history society (d) Bombay natural habitat society
63. MoEF is known as:
(a) monitoring of environment and forest (b) ministry of environment and forests
(c) municipality of environmental forum (d) none of the above
64. Mega hydro projects are harmful due to:
(a) security (b) natural calamities
(c) flood (d) all
65. Maximum oxygen production habitat is:
(a) forest (b) urban
(c) cities (d) rivers
66. Dams are beneficial for:
(a) irrigation (b) drinking water supply
(c) fish farming (d) all
67. The Indian National River is:
(a) Ganges (b) Krishna
(c) Tista (d) All
68. To solve the water problem, the technique used is known as:
(a) interlinking the rivers (b) construction of the dams
(c) rivers (d) all
69. The major purpose of most of the dams around the world is:
(a) electricity (b) irrigation
(c) flood control (d) all
70. Major disadvantages of dams are:
(a) flood (b) rehabilitation
(c) environmental problems (d) all
71. In a food chain, the total amount of living material is depicted by:
(a) pyramid of energy (b) pyramid of biomass
(c) pyramid of numbers (d) all of these
72. The two types of energy flow are:
(a) Z-Y Shape (b) X-Y Shape
(c) line-Y Shape (d) A-B Shape
73. Which one is the correct food chain?
(a) Phytoplankton→Zooplankton→Fish (b) Grass→Fish→Zooplankton
(c) Zooplankton→Phytoplankton→Smallfish (d) Protozoans→Algae→Zooplankton
74. Which group of vertebrates comprises the highest number of endangered species?
(a) Birds (b) Mammals
(c) Reptiles (d) Fishes

75. Which of the following is an *in situ* conservation measure taken by India?
(a) Project Tiger (b) Project Elephant
(c) Project Rhino (d) All

76. Which area is rich in biodiversity?
(a) Silent valley (b) NDBR
(c) Ranthambor (d) Datchigam

77. Symbol of WWF is:
(a) Red Panda (b) Giant Panda
(c) Black Panda (d) Teddy Panda

78. Who decided the protected areas
(a) WCMC (b) UNEP
(c) MoEF (d) BMC

79. Greenhouse gases are:
(a) CH_4 CO_2 (b) NO_2 CO_2
(c) CO CO_2 (d) SO_2 NO_2

80. Mercury is responsible for:
(a) Minamata (b) saphogation
(c) dehydration (d) all

81. Noise pollution can be controlled
(a) plantation (b) ears plugs
(c) sound proof instruments (d) none

82. The main source of soil pollution is:
(a) chemical fertilizer (b) pesticides
(c) herbicides (d) all

83. Ecological effects of thermal pollution are:
(a) decreased DO level (b) decreased CO_2 level
(c) decreased temperature (d) all

84. PAN means:
(a) Para Acetyl Nitrate (b) Para Amide Nitrate
(c) Both (d) None

85. Common energy sources in Indian villages are:
(a) electricity (b) coal
(c) sun (d) wood and animal dung

86. Drought is a:
(a) natural disaster (b) man-made calamities
(c) none of these (d) both

87. Hydroelectric energy is generated by:
(a) kinetic energy of water (b) kinetic energy of wind
(c) kinetic energy of aquatic life (d) none

88. Biodiesel is produced from:
(a) vegetable oils (b) animal fats
(c) both (d) animal fats only

89. Kaziranga National Park is famous for:
(a) tiger (b) elephant
(c) rhinoceros (d) musk deer

90. National Parks come under:
(a) *In situ* (b) *Ex situ*
(c) Zoo (d) Biosphere reserve

91. Which is the permissible range of pH for drinking water as per the Indian standards?
(a) 6 to 9 (b) 6.5 to 7.5
(c) 6 to 8.5 (d) 7.5 to 9.5

92. The function of ecosystem is:
(a) flow of chemical nutrients (b) flow of various gases
(c) flow of water (d) none

93. Global atmospheric temperature is likely to increased due to:
(a) burning of fossil fuel (b) water pollution
(c) soil erosion (d) none of these

94. Tropical rainforests occur in India in:
(a) Jammu and Kashmir (b) Rajasthan
(c) Kerala and Assam (d) nowhere

95. Chipko movement was started by:
(a) Chandi Prasad Bhatt (b) Sunderlal Bahuguna
(c) Gaura Devi (d) None of them

96. Minerals essential for the economy of a nation are:
(a) strategic minerals (b) critical minerals
(c) energy generating minerals only (d) none of these

97. The overnourished lakes with 'algal blooms' are called:
(a) eutrophic (b) oligotrophic
(c) dystrophic (d) meromictic

98. Polyethylene tetraphalate (PET) is found in:
(a) used motor vehicle oil (b) soft drink bottles
(c) glass (d) steel cars

99. Who used the term 'biodiversity crises' for the first time?
(a) R.H. Whittker (b) E.O Wilson
(c) R.D. Barnes (d) N. Mayers

100. Tsunami is a:
(a) Chinese term (b) Japanese term
(c) Indian term (d) None of these

ANSWERS

1. (c)	2. (d)	3. (c)	4. (b)	5. (d)	6. (c)	7. (b)	8. (d)	9. (c)	10. (b)
11. (b)	12. (d)	13. (b)	14. (a)	15. (c)	16. (c)	17. (b)	18. (c)	19. (b)	20. (d)
21. (c)	22. (a)	23. (b)	24. (b)	25. (c)	26. (d)	27. (c)	28. (a)	29. (d)	30. (c)
31. (b)	32. (b)	33. (c)	34. (b)	35. (b)	36. (d)	37. (c)	38. (a)	39. (c)	40. (a)
41. (d)	42. (a)	43. (d)	44. (a)	45. (c)	46. (c)	47. (b)	48. (a)	49. (a)	50. (d)
51. (a)	52. (a)	53. (c)	54. (a)	55. (a)	56. (b)	57. (d)	58. (a)	59. (c)	60. (d)
61. (c)	62. (b)	63. (b)	64. (d)	65. (d)	66. (d)	67. (a)	68. (a)	69. (a)	70. (b)
71. (c)	72. (c)	73. (a)	74. (d)	75. (a)	76. (a)	77. (b)	78. (a)	79. (d)	80. (a)
81. (a)	82. (d)	83. (a)	84. (a)	85. (d)	86. (a)	87. (a)	88. (c)	89. (c)	90. (a)
91. (b)	92. (a)	93. (a)	94. (c)	95. (c)	96. (c)	97. (a)	98. (b)	99. (b)	100. (b)

IMPORTANT OBJECTIVE TYPE QUESTIONS

1. Chipko movement was started in:
   (a) 1972 (b) 1973
   (c) 1980 (d) 1974
2. Total biosphere reserves in India are:
   (a) 19 (b) 17
   (c) 18 (d) 16
3. Chemical smog is also known as:
   (a) London smog (b) Britain smog
   (c) Australia smog (d) India smog
4. Smog is the combination of:
   (a) Fog and smoke (b) SO_2 and CO
   (c) Fog and SO_2 (d) Smoke and SO_2
5. The term ecosystem was given by:
   (a) Warming (b) Clements
   (c) Tansley (d) None
6. Artificial ecosystem is:
   (a) aquarium (b) forest
   (c) grassland (d) cropland
7. Plants growing on sandy soil are called:
   (a) psammophytes (b) lithophytes
   (c) xerophytes (d) none
8. The term ecological succession was given by:
   (a) Warming (b) Hult
   (c) Tansley (d) None
9. When succession starts on bare rocks it is called:
   (a) psammosere (b) xerosere
   (c) lithorsere (d) none
10. Mutualism is:
    (a) – + (b) – –
    (c) + 0 (d) + +
11. The reservoir for nitrogen is:
    (a) atmosphere (b) rocks
    (c) soil (d) water
12. Bacteria drives the ……cycle.
    (a) nitrogen (b) energy
    (c) water (d) oxygen

13. Higher plants most often absorb nitrogen from the soil in the form of:
    (a) nitrogen (b) nitrites
    (c) ammonia (d) amino acids
14. When the vegetation of any region reaches climatic climax it is:
    (a) lithophytic (b) mesophytic
    (c) xerophytic (d) hydrophytic
15. The importance of ecosystem lies in:
    (a) flow of energy (b) Cycling of materials
    (c) Both (d) None
16. Food levels of an ecosystem are called:
    (a) consumer levels (b) producer levels
    (c) trophic levels (d) none
17. A lake ecosystem is:
    (a) artificial (b) abiotic
    (c) a & b (d) natural
18. Pyramid of energy in a pond ecosystem is always:
    (a) inverted (b) irregular
    (c) linear (d) upright
19. Man in ecosystem is:
    (a) producer (b) primary consumer
    (c) secondary consumer (d) both b and c
20. The ultimate trophic level in a food chain is occupied by:
    (a) decomposers (b) top consumers
    (c) tertiary consumers (d) producers
21. In a predator (grazing) food chain
    (a) organic food is dead matter like fallen leaves
    (b) organic food is obtained by killing the prey
    (c) organic food is manufactured by autotrophs
    (d) food is taken from host
22. The stability of a grassland ecosystem depends upon:
    (a) carnivores (b) grazers
    (c) both (d) none
23. Savannahs are:
    (a) tropical rainforests (b) desert scrubs
    (c) grasslands with scattered trees (d) dense forests
24. A habitat with long severe winter and growing season of few months of summer form:
    (a) grasslands (b) tundra ecosystem
    (c) taiga (d) savannah

25. The type of food chain in which organic matter is decomposed by microbes and converted into energy rich compounds is called:
    (a) detritus food chain (b) grazing food chain
    (c) cybernetics (d) none
26. In a food chain, the largest population is that of:
    (a) primary consumers (b) secondary consumers
    (c) producers (d) decomposers
27. Pyramid of numbers is upright in:
    (a) grassland ecosystem (b) tree ecosystem
    (c) pond ecosystem (d) none of these
28. The shape of pyramid of numbers in a pond ecosystem is.
    (a) linear (b) spindle
    (c) upright (d) inverted
29. Which of the following has the maximum biomass?
    (a) Temperate forest (b) Tropical forests
    (c) Alpine vegetation (d) Taiga
30. Red Data Book refers to.
    (a) a book with red cover (b) a book with a list of extinct species
    (c) a book a list of dead species (d) book written by Karl Marx
31. Total biogeographical zones in India are:
    (a) 5 (b) 7
    (c) 12 (d)10
32. Hot spot area refers to:
    (a) hot climate area (b) endemic species area
    (c) endangered species area (d) rare species area
33. Who is the Union Minister of Environment and Forestry?
    (a) Mr. Jai Ram (b) Dr. Manmohan Singh
    (c) Mr. T.R Balu (d) Mr. T.R. Sankarraman
34. Who is the Environment and Forest Minister of Uttarakhand?
    (a) Mr. Bansidhar Bhagat (b) Mr. Bisan S. Chaupal
    (c) Dr. Ramesh Pokhriyal (d) Mr. Madan Kausik
35. CITES stands for:
    (a) Conservation for Indian Tiger and Ecosystems
    (b) Conservation on International Trade in Endangered Species
    (c) Convention on International Trade in Endangered Species
36. Wildlife Protection Act was passed in:
    (a) 1971 (b) 1972
    (c) 1978 (d) 1974

37. Project Tiger was launched in:
    (a) 1965 (b) 1973
    (c) 1972 (d) 1975
38. World Biodiversity day is celebrated on:
    (a) 12 May (b) 23 May
    (c) 24 May (d) 22 May
39. Total protected area in India is:
    (a) 112 km^2 (b) 581 km^2
    (c) 584 km^2 (d) 589 km^2
40. Biodiversity Act was passed in India in:
    (a) 2000 (b) 2001
    (c) 2002 (d) Still waiting
41. Which on the following species has become extinct?
    (a) Dodo (b) Red headed vulture
    (c) Eagle (d) Hawk
42. IUCN refers to:
    (a) International Union for Conservation Nature
    (b) International Union for Conservation Nature and Natural Resources
    (c) International Union for Conserving Nature
43. UNEP refers to:
    (a) United Nation Environmental Project
    (b) United Nation Environment Programme
    (c) United Nation Environment Protection
    (d) United Nation Environment Protection Act
44. Which one is not a wildlife sanctuary?
    (a) Motichur Wildlife Sanctuary
    (b) Binsar Wildlife Sanctuary
    (c) Naina Wildlife Sanctuary
    (d) Nanda Devi Wildlife Sanctuary
45. How many hot spot areas have been introduced:
    (a) 12 (b) 24
    (c) 34 (d) 39
46. Indian biodiversity stands at which place at global level?
    (a) 1 (b) 12
    (c) 10 (d) 20
47. National parks come under:
    (a) *In situ* (b) *Ex situ*
    (c) Zoo (d) Biosphere reserve

48. Biosphere refers to:
(a) Conserve the wildlife (b) Conserve the birds
(c) Conserve the vegetation (d) Conserve the ecosystem

49. How many wetlands have been introduced in India?
(a) 12 (b) 15
(c) 25 (d) 10

50. Wetland sites are also known as:
(a) Rem sites (b) Ramsar sites
(c) Ramsor sites (d) Ramsey sites

51. How many biogeographical zones are there in the world
(a) 10 (b) 12
(c) 14 (d) 20

52. What is the state flower of Uttarakhand?
(a) Lotus (b) Burans
(c) Bram-kamal (d) Rose

53. The name of biodiversity was introduced in:
(a) 1990 (b) 1992
(c) 1994 (d) 2001

54. How can we conserve our biodiversity?
(a) Through awareness (b) Through indigenous knowledge
(c) Policies and laws (d) Through education

55. Which is a non-metallic mineral?
(a) Cu (b) Coal
(c) Fe (d) Zn

56. Kolar mine is related to:
(a) Pt (b) Au
(c) Fe (d) Cu

57. Panna is situated in which state?
(a) Rajasthan (b) MP
(c) UP (d) UK

58. In which state maximum coal occurs?
(a) UP (b) UK
(c) MP (d) Orissa

59. A drought-prone area programme was launched in:
(a) 1979 (b) 1988
(c) 1985 (d) 1995

60. Which mineral occurs in the chlorophyll?
(a) Mg (b) Cl
(c) Cu (d) Fe

61. BMW stands for:
(a) British Motors Works (b) Board of Mineral Water
(c) Bio Medical Waste (d) Biological Medical Waste
62. Study of natural calamities is known as:
(a) disaster (b) earthquake
(c) flood (d) forest fire
63. Where is large wind farm of our country situated?
(a) Kanyakumari (b) Lucknow
(c) Dehradun (d) Shimla
64. Burning of coal produces pollution of:
(a) NO_2 (b) SO_2
(c) O_3 (d) None
65. The largest amount of ozone is found in which part of atmosphere:
(a) upper stratosphere (b) upper mesosphere
(c) upper troposphere (d) lithosphere
66. The only gas in the atmosphere that can absorb the sun's dangerous UV is:
(a) nitrogen (b) oxygen
(c) hydrogen (d) ozone
67. How much % of methane occurs in natural gas?
(a) 80% (b) 85%
(c) 95% (d) 70%
68. Which is the secondary air pollutant.
(a) CO_2 (b) SO_2
(c) O_3 (d) NO_2
69. Unit of ozone measurement is:
(a) ppm (b) Domg/l
(c) Dobson (d) Decibel
70. Which light range ozone absorbs?
(a) UV light (b) B-UV light
(c) C-UV light (d) None
71. The cause of food problem is:
(a) insufficient production (b) black marketing
(c) population (d) none
72. Air pollution is not caused by:
(a) pollen grain (b) hydroelectric power
(c) industries (d) all
73. Ashok and neem plants are planted in front of buildings because they:
(a) absorb harmful gases (b) absorb noise
(c) emit scent (d) none

74. The impact of drought is:
(a) increased water level
(b) population decrease
(c) decreased fertility
(d) increased water-borne disease

75. The causes of water pollution are:
(a) automobiles industry
(b) deforestation
(c) DDT
(d) none

76. Earth submit was held at:
(a) Reo-de-Jinario
(b) USA
(c) India
(d) Nepal

77. Taj Mahal is threatend due to:
(a) SO_2
(b) HC
(c) Cl
(d) all

78. Air pollution effects are usually seen on:
(a) leaves
(b) flower
(c) stem
(d) root

79. World ozone day is on:
(a) 30 Jan
(b) 21 Sept
(c) 16 Sept
(d) None

80. The basic component of smog is:
(a) Ozone
(b) Ozone + PAN
(c) PAN
(d) DDT

81. Acid rain occurs due to:
(a) SO_2
(b) CO
(c) Cl
(d) none

82. Maximum percentage of acid in acid rain is:
(a) HNO_3
(b) HCl
(c) H_2SO_4
(d) CH_3 COOH

83. Classical smog is also known as:
(a) Los Angeles Smog
(b) London Smog
(c) Tokyo Smog
(d) none

84. Los Angeles smog occurs due to:
(a) $SO_2 + H_2S$
(b) $O_3 + NO_2$
(c) $CO_2 + CO$
(d) none

85. Audible sound has an intensity of:
(a) –2db
(b) –5db
(c) 0db
(d) –10db

86. Blue Baby syndrome is caused due to:
(a) nitrate
(b) CO_2
(c) sulphate
(d) Hg

87. Radioactive disaster occurred in:
(a) 1945 (b) 1950
(c) 1952 (d) 2010

88. How many hot spot regions are there in the world:
(a) 37 (b) 39
(c) 36 (d) 35

89. First National Park was:
(a) Rajaji (b) Jim Corbett
(c) Kanha (d) Dudwa

90. Wild ass occurs in:
(a) Runn of Kutch (b) Giri
(c) Periyar (d) None

91. Producers are also known as:
(a) autotrophs (b) chemotrophs
(c) heterotrophs (d) none

92. Rajiv Gandhi National Park is located in:
(a) Gujarat (b) Rajasthan
(c) Tamil Nadu (d) Uttar Pradesh

93. Free-living nitrogen fixing bacteria is:
(a) *Rhizobium* (b) *Azotobacter*
(c) *Penicillium* (d) *Nitrosomonas*

94. Plants growing on saline soil are called:
(a) helophytes (b) halophytes
(c) lithophytes (d) samophytes

95. *Ex situ* conservation example is:
(a) gene bank (b) botanical garden
(c) seed banks (d) all

96. *In situ* conservation is:
(a) national park (b) sanctuaries
(c) wetland (d) all

97. Who started the Chippko Movement in India?
(a) Gaura devi (b) Gayatri devi
(c) Gaurangi Devi (d) Gaoru Devi

98. Following is always an upward pyramid:
(a) energy (b) biomass
(c) numbers (d) all

99. World Earth Summit 2009 was based on:
(a) air pollution (b) global warming
(c) wildlife protection (d) none

100. Poor water resources countries are located on:
   (a) sediment rocks (b) volcanic rocks
   (c) igneous & meta morphic rocks (d) all
101. Decomposers are:
   (a) decompose organic matter (b) microorganisms
   (c) saprophytes (d) all
102. Examples of macrophytes aquatic producers are:
   (a) Hydrilla (b) Typha
   (c) Eutricularia (d) All
103. Which one of the following disasters is maximum harmful?
   (a) flood (b) earthquake
   (c) fire (d) drought
104. How much percentage of total solar energy is utilized by the plants for photosynthesis:
   (a) 1–4% (b) 5%
   (c) 4.1% (d) 5.1%

ANSWERS

1. (b)	2. (d)	3. (a)	4. (a)	5. (c)	6. (a)	7. (a)	8. (a)	9. (b)	10. (d)
11. (a)	12. (c)	13. (d)	14. (d)	15. (d)	16. (c)	17. (b)	18. (b)	19. (c)	20. (b)
21. (a)	22. (c)	23. (c)	24. (c)	25. (b)	26. (b)	27. (c)	28. (b)	29. (a)	30. (c)
31. (b)	32. (b)	33. (c)	34. (b)	35. (d)	36. (c)	37. (a)	38. (b)	39. (b)	40. (c)
41. (d)	42. (c)	43. (a)	44. (d)	45. (b)	46. (b)	47. (b)	48. (c)	49. (b)	50. (b)
51. (b)	52. (b)	53. (b)	54. (d)	55. (b)	56. (a)	57. (d)	58. (a)	59. (a)	60. (b)
61. (a)	62. (d)	63. (a)	64. (c)	65. (c)	66. (b)	67. (c)	68. (b)	69. (b)	70. (c)
71. (c)	72. (a)	73. (a)	74. (a)	75. (c)	76. (c)	77. (a)	78. (c)	79. (b)	80. (c)
81. (c)	82. (d)	83. (a)	84. (b)	85. (b)	86. (a)	87. (a)	88. (b)	89. (d)	90. (d)
91. (d)	92. (d)	93. (a)	94. (a)	95. (b)	96. (b)	97. (c)	98. (d)	99. (a)	100. (b)
101. (c)	102. (a)	103. (a)	104 (b)						

Class Test for B. Tech Students
ENVIRONMENTAL STUDIES

Total Marks: 50

1. Kyoto protocol is related to:
   (a) global warming (b) ozone depletion
   (c) biodiversity (d) all of these
2. Which one is a major global warming gas?
   (a) Water vapor (b) Ethane
   (c) Carbon monoxide (d) All of these
3. Major component of acid rain is:
   (a) H_2SO_4 (b) HCl
   (c) CH_3COOH (d) HCOOH
4. Biofuel is another source of traditional energy, which is produced from:
   (a) Jatropha (b) Banyan
   (c) Neem (d) Bamboo
5. Blue Baby syndrome develops in infants due to:
   (a) excess of cadmium (b) fluoride
   (c) arsenic (d) nitrate in drinking water
6. Which one of the following is not a greenhouse gas?
   (a) Water vapour (b) Halogen compound
   (c) CH_4 (d) SO_2
7. The main sources of biofertilizers are:
   (a) bacteria (b) fungi
   (c) cynobacteria (d) all
8. The chief source of nitrogen for plants in soil is:
   (a) nitrite (b) nitrate
   (c) free nitrogen (d) none
9. Wasteland sites can be used for:
   (a) aforestation (b) agriculture
   (c) fish farming (d) none
10. Deforestation will decrease by:
    (a) rainfall (b) soil erosion
    (c) cyclones (d) land slides
11. Air pollution is caused by:
    (a) automobile exhausts (b) radioactive substance
    (c) smoke (d) all the above
12. In cities like Bombay and Culcutta, the major air pollutants are:
    (a) carbon monoxide and oxides of sulphur

(b) hydrocarbons and hot air
(c) algal spores and marsh gas
(d) ozone.

13. Which is not usually an air pollutant?
(a) CO_2 (b) SO_2
(c) Hydrocarbons (d) NO

14. Lichens are main indicator of:
(a) water pollution (b) SO_2 pollution
(c) radioactive pollution (d) soil pollution

15. Minamata disease is caused by:
(a) Hg pollution (b) Pb pollution
(c) sewage pollution (d) greenhouse effect

16. Which of the following is not a non-biodegradable substance?
(a) Sewage (b) DDT
(c) Aldrin (d) Chlorinated hydrocarbons

17. Which is a secondary air pollutant?
(a) CO_2 (b) SO_2
(c) NO_2 (d) O_3

18. Impacts of global warming are:
(a) hurricanes (b) floods
(c) marine diseases (d) all of the above

19. The treatment method to reduce fluorides is:
(a) reverse osmosis (b) filtration
(c) areation (d) boiling

20. Shells of corals are formed by:
(a) calcium (b) nitrogen
(c) carbon (d) none

21. The conversion of nitrate to ammonia is called:
(a) nitrification (b) ammonification
(c) nitrogen fixation (d) denitrification

22. When BOD is high, the dissolved oxygen content of water becomes
(a) low (b) high
(c) medium (d) none

23. Among all the natural disasters, largest number of people in the world is affected by:
(a) cyclone (b) floods
(c) drought (d) earthquakes

24. Gold is produced from:
(a) quartz (b) laterite
(c) graphite (d) none

25. The excessive nutrition due to fertilizers causes:
    (a) cholera (b) vomiting
    (c) Blue Baby syndromes (d) Eutrophication
26. The water control methods include:
    (a) flood proofing (b) flood warning
    (c) catchment modification (d) both a & c
27. Where does tropical rain forest occur in India?
    (a) Assam (b) Himachal Pradesh
    (c) Tamil Nadu (d) Orissa
28. The domestic burning of coal and other fuels produces mainly:
    (a) SPM (b) carbon dioxide
    (c) explosive fumes (d) all
29. The amount of oxygen required for biochemical decomposition processes is called:
    (a) dissolved oxygen (b) BOD
    (c) COD (d) none
30. Maximum SO_2 pollution occurs in which city?
    (a) Delhi (b) Bombay
    (c) Culcutta (d) Kanpur
31. Who defined pollution?
    (a) Tansely (b) Darwin
    (c) Odum (d) Ramen
32. Which one is the secondary air pollutant?
    (a) O_3 (b) CO
    (c) HF (d) SO_2
33. SO_2 enters into plant.:
    (a) stem (b) root
    (c) flower (d) stomata
34. Conservative pollutants are:
    (a) glass (b) long chain phenolic compound
    (c) DDT (d) all
35. Which of the following acts as a competitive inhibitor of CO_2 on PEP-carboxylase?
    (a) SO_2 (b) NO_2
    (c) O_3 (d) CO
36. Which of the following checks the hill reaction of photosynthesis?
    (a) PAN (b) O_3
    (c) CO (d) Pb
37. The Earth Summit of Rio-de Janeiro was held in the year:
    (a) 1992 (b) 1993
    (c) 1993 (d) 1994

38. The main component of classical smog is:
(a) PAN (b) O_3
(c) NO_2 (d) SO_2
39. Green Revolution means:
(a) Increase in production of food plants.
(b) increase in growth of green plants for maintaining ecosystem balance
(c) Growth of green plants in order to check soil erosion
(d) None of the above
40. Genetic diversity is shown by the plants due to:
(a) intraspecific breeding (b) interspecific breeding
(c) both a & b (d) none
41. Maintenance of genetic diversity in natural reserves like national parks is called
(a) *ex situ* conservation (b) *in situ* conservation
(c) gene pool (d) none
42. Who is known as the Father of Green Revolution?
(a) Vavilov (b) N. Borlaug
(c) Swaminathan (d) G.H. Khush
43. The ultimate source of energy for an ecosystem is:
(a) sunlight (b) ATP
(c) DNA (d) water current
44. LPG cooking gas is:
(a) low pressure gas (b) biogas
(c) fossil fuel (d) low price gas
45. The energy produced by hydel-power plant is:
(a) renewable and polluting (b) non-renewable and non-polluting
(c) renewable and non-polluting (d) none
46. The percentage of methane in biogas is:
(a) 70 (b) 90
(c) 7 (d) 20
47. Non-renewable substance or resource is:
(a) planktons and fishes (b) fossil and minerals
(c) animals and plants (d) animals and minerals
48. Biofertilizers are:
(a) Cow dung manure and farmyard waste (b) Aquik growing crop ploughed under
(c) *Anabaena and Azolla* (d) None
49. The natural heat of earth is called:
(a) abiotic energy (b) bio energy
(c) geothermal energy (d) geographic energy

50. Which of the following source of energy is non-renewable?
    (a) Wildlife
    (b) Forest wealth
    (c) Coal reserves
    (d) Hydel power plant

ANSWERS

1. (b)	2. (a)	3. (b)	4. (d)	5. (b)	6. (a)	7. (a)	8. (c)	9. (c)	10. (c)
11. (d)	12. (d)	13. (c)	14. (a)	15. (d)	16. (c)	17. (a)	18. (b)	19. (d)	20. (b)
21. (a)	22. (d)	23. (a)	24. (b)	25. (a)	26. (c)	27. (b)	28. (c)	29. (b)	30. (a)
31. (d)	32. (a)	33. (b)	34. (c)	35. (b)	36. (d)	37. (b)	38. (c)	39. (b)	40. (c)
41. (a)	42. (c)	43. (a)	44. (a)	45. (d)	46. (d)	47. (a)	48. (a)	49. (c)	50. (c)

ABBREVIATIONS

BSI	:	Botanical Survey of India
CAZRI	:	Central Arid Zone Research Institute
CFC	:	Chlorofluorocarbons
CITES	:	Conservation in International Trade in Endangered Species
CPCB	:	Central Pollution Control Board
EE	:	Environment and Ecology
EIA	:	Environment Impact Assessment
ENVIS	:	Environmental Information System
EPA	:	Environment Protection Act
WII	:	Wildlife Institute of India
UNDP	:	United Nation Development Programme
UNEP	:	United Nation Environment Programme
WWF	:	World Wildlife Fund
BNHS	:	Bombay Natural History Society
IUCN	:	International Union for Conservation of Nature
MAB	:	Man and Biosphere
NEPA	:	National Environmental Policy Act
ZSI	:	Zoological Survey of India
UNESCO	:	United Nations Educational, Scientific and Culture Organization
NWDB	:	National Wasteland Development Board
MoEF	:	Ministry of Environment and Forestry
NEERI	:	National Environmental Engineering Research Institute
TERI	:	Tata Energy Research Institute
BOD	:	Biological Oxygen Demand
COD	:	Chemical Oxygen Demand
EMP	:	Environment Management Plan
WCMC	:	World Conservation Monitoring Centre
CITES	:	Convention on International Trade in Endangered Species
WTI	:	Wildlife Trust of India

IMPORTANT DAYS WITH REFERENCE TO ENVIRONMENT

World Environment Day	:	5th June
World Population Day	:	11st July
World Forest Day	:	21st March
National Science Day	:	28th February
Earth Day	:	22nd April
Wildlife Week	:	First week of October
Ozone Day	:	16th September
World Sparrow Day	:	20th March
World Wetland Day	:	2nd February
National Elephant Day	:	13 rd March
International Biodiversity Day	:	22nd May
World Water Day	:	22 March

REFERENCES

Barnes HL, 1989. Cross-Generational Coalitions, Discrepant Perceptions and family functioning. In David H, Ols on, Candyce S. Russell, Douglas H, Sprenkle ed. Cricunplese Model Systemic. Assessment and Treatment of families (pp. 175-198). Binghamton, New York; The Howorth Press.

Barthlott W, Lauer W, Placki A (1996) Global distribution of species diversity in vascular plants: Towards a world map of Phytodiversity. *Erdkunde* 50: 317–327.

CSE, 1991. Floods, Flood Plains and Environmental Myths., Center for Science and Environment, New Delhi.

Dhar ON, and Nandargi S, 1998. Flood studies in India, In Kale, VS (Ed) Geological Society of India Bangalore Mencoir, 4: 53.

Groombridge B, Jenkins MD (2000) Global Biodiversity: Earth's living sources in the 21st century UK: World Conservation Monitoring Committee. 246.

Heywood VH, Watson RT (1995) Global Biodiversity Assessment UNEP. UK: Cambridge University Press.

Kale VS, 1999, Joul. Geol. Soc. Ind. 53: 5.

Khoshoo TN (1996) Biodiversity in the India Himalayas: conservation and utilization. In: Sheggi P, ed. Banking on Biodiversity. Kathmandu: International Centre for Integrated Mountain Development.

Kumar HD, Forest Resources; Conservation and Management, Affiliated East West Press Pvt. Ltd., 2001.

May RM, 1990. How many species ? Phil-Trans. R. Sol. Lond. B 330: 293-304.

Mayer *et al.*, 1979; Mayer, 1988; Barnes, 1989; May, 1990

McNeely JA, Miller KR, Reid WV, Mittameier RA, Werner TB (1990) Conserving the world's Biological Diversity Washington, D.C. and Gland, Switzerland: WRI, World Conservation Union, World Bank WWF-US and conservation International.

MoEF, 2000. National Policy and Macro level Action & Strategy on Biodiversity New Delhi: Ministry of Environment and Forests, Government of India. 74.

Myers N (1992) Future Operational Monitoring of Tropical Forests: An Alert Strategy. Joint research centre, commission of the European Community, Ispra, Italy.

Myers N (1988). Threatened biotr! 'hot spot' in tropical forests Environmentalist 8: 187-208.

Navneet Rai (2011) Role of Emulsifiers & Demulsifiers in Various Emulsions, Microemulsions & Nanoemulsions with Their Applications Ph.D. Thesis, H.N.B. Garhwal University.

Rodgers WA, Panwar HS, Mathur VB (2000) Wildlife Orotection Area Network in India: A Review (Executive Summary) Dehradun: Wildlife Institute of India 44.

Solbrig OT, 1993. All of US-1-Sept.

Solbrig, 1993

www.biodiversity.com

INDEX